The Changing Population of Europe

# The Changing Population of Europe

Edited by
Daniel Noin and Robert Woods

First published 1993

Blackwell Publishers
108 Cowley Road
Oxford OX4 1JF
UK

238 Main Street, Suite 501
Cambridge, Massachusetts 02142
USA



*British Library Cataloguing in Publication Data*

A CIP catalogue record for this book is available from the British Library.

*Library of Congress Cataloging-in-Publication Data*

The Changing population of Europe/edited by Daniel Noin & Robert Woods.
p. cm.
Includes bibliographical references and index.
ISBN 0-631-17635-7 (alk. paper). – ISBN 0-631-18972-6 (pbk. : alk. paper)
1. European Economic Community countries – Population.
2. Population forecasting–European Economic Community countries.
I. Noin, Daniel. II. Woods, Robert.
HB3582.5.C48 1993
304.6′094 – dc20 92-430003
CIP

Typeset in Sabon on 10/12 pt
by Best-set Typesetter Ltd., Hong Kong
Printed in Great Britain by Hartnolls Ltd, Bodmin, Cornwall

This book is printed on acid-free paper

# Contents

# List of Figures

# List of Tables

# List of Contributors

Bähr, Jürgen
Professor of Geography, Department of Geography, University of Kiel, Germany

Brunetta, Giovanna
Professor of Geography, Department of Geography, University of Padua, Italy

Champion, A. G.
Senior Lecturer in Geography, Department of Geography, University of Newcastle upon Tyne, England

Garcia Ballesteros, Aurora
Professor of Geography, Department of Geography, Complutense University, Madrid, Spain

Hall, Ray
Senior Lecturer in Geography, Department of Geography and Earth Sciences, Queen Mary and Westfield College, University of London, England

Hecht, Jacqueline
Researcher, Institut National d'Etudes Démographiques, Paris, France

Köhli, Jörg
Researcher, Department of Geography, University of Kiel, Germany

Leridon, Henri
Chef du Département de Socio-démographie, Institut National d'Etudes Démographiques, Paris, France

Noin, Daniel
Professor of Geography, Department of Geography, University of Paris 1 (Sorbonne), France

Ogden, Philip E.
Reader in Geography, Department of Geography and Earth Sciences, Queen Mary and Westfield College, University of London, England
Renard, Jean
Professor of Geography, Department of Geography, University of Lille, France
Salt, John
Senior Lecturer in Geography, Department of Geography, University College, University of London, England
Simon, Gildas
Professor of Geography, Department of Geography, University of Poitiers, France
Sporton, Deborah
Lecturer in Geography, Department of Geography, University of Sheffield, England
Thumerelle, Pierre-Jean
Professor of Geography, Department of Geography, University of Lille, France
Vandermotten, Christian
Professor of Geography, Department of Geography, Free University of Brussels, Belgium
Warnes, Anthony M.
Professor of Geography, Department of Geography, King's College, University of London, England
White, Paul
Senior Lecturer in Geography, Department of Geography, University of Sheffield, England
Woods, Robert
Professor of Geography and Director of the Graduate Programme in Population Studies, Department of Geography, University of Liverpool, England

# Preface

The characteristics of populations are very often at the heart of social issues and demographic change has always been a fundamental element of the evolution of nation states. It is therefore essential to study these particular aspects of change very carefully. However, among the many published books and articles on the European Community (EC), those concentrating on its population are relatively few and far between.

Although we are not exclusively concerned with the population of the EC, it is our principal focus. The Nordic countries, Switzerland, Austria and the states of eastern Europe are not ignored, but they are given second place in the ordering of our priorities. This is of course to be regretted, but it may be rectified in the future by the preparation of a complementary volume which takes as its theme the population of eastern Europe.

This work, prepared by a team of European geographers and demographers, aims to fill that gap. It seeks to answer those questions raised by the geo-demographic study of the Community: what are the principal features of the Community's population in the early 1990s? How do they vary spatially? How have their demographic characteristics evolved during the decades since 1945? What are the most distinctive recent trends? What is the outlook for the twenty-first century?

Finally, the editors would like to express their thanks to Sandra Mather for drawing the maps and most of the diagrams, to Chris Galley for compiling the index and to the contributors for their good humour and patience.

Robert Woods, *University of Liverpool*
Daniel Noin, *Université de Paris I*

# Abbreviations

We have used certain conventions in editing this book and these should be noted at the outset. The European Community is often referred to simply as the Community or the EC, and sometimes the EC of 12 etc. Germany (D) refers to the reunified state and prior to 1990 the Federal Republic of Germany or West Germany (FRG) is distinguished from the German Democratic Republic or East Germany (GDR). The Republic of Ireland is consistently referred to as Eire (IRL). While the United Kingdom (UK) comprises England, Wales, Scotland and Northern Ireland, Great Britain (GB) is made up of England and Wales and Scotland. The following abbreviations are also used on occasions: Denmark, DK; the Netherlands, NL; Belgium, B; Luxembourg, L; France, F; Spain, SP or E; Portugal, P; Italy, I; Greece, GR; and the Commonwealth of Independent States, CIS. The former Soviet Union or USSR is occasionally used.

In common with most disciplines, demography and geography have their own technical terms and definitions which are often expressed in abbreviated form. CBR and CDR are used for, respectively, the crude birth and death rates which are both expressed in parts per thousand. IMR stands for the infant mortality rate in parts per thousand live births. Life expectancy at birth is given in years and represents the average length of life that a new-born baby might expect. TFR stands for the total fertility rate. It expresses the average number of live born babies that a woman may be expected to have born by the time she reaches the menopause. MHS stands for mean household size.

# 1

# The European Community and its Population

Daniel Noin

The European Community (EC) is a commonwealth of states which since the beginning of 1993 has constituted a single market allowing for the free circulation of people, goods and capital. For many Europeans, this date marked a point when they finally became genuine European citizens; moreover, many possess the same bordeaux-coloured passport and believe that the old frontiers have practically disappeared. This is more a dream than reality. It is true that there is already a solid 'economic Europe' with numerous and growing links between member countries. There is also undoubtedly a Europe of students, of researchers, of artists, of business people and of company executives. However, there is no 'political Europe'. In the case of a serious international crisis, Europe would not speak with a single voice. The different states of the Community have learned to co-operate in all sorts of areas, though, until now at least, they have hardly abandoned even tiny parcels of sovereignty. It is surely not impossible that a true union will be formed with common defence and foreign policies. This has been long-awaited though it still seems a far way off.

The countries of the Community have nevertheless come a long way over the last 30 years. Europe is slowly advancing on the path to unity; little by little, new institutions and commonly adopted guidelines are contributing to a new Community state of mind among both leaders and ordinary citizens.

This commonwealth of states has not only experienced the strengthening of existing links between its members, but an expansion too. The initial group, formed after the signing of the

Treaty of Rome in 1957, comprised six states: Belgium, France, Italy, Luxembourg, the Netherlands and the Federal Republic of Germany. It has been enlarged into a community of 12 states with the addition of Denmark, Eire and the UK in 1973, of Greece in 1981 and finally of Spain and Portugal in 1986. Moreover, the former German Democratic Republic also joined in 1990 following German reunification. Although it is perhaps desirable to take a pause in this process of rapid growth, it is possible that other states, notably Austria and Sweden, will be admitted in years to come. It is also certain that other countries that are not now formal members will strengthen their economic links, thus moving along a road that may ultimately lead to a single federal state of Europe.

### A Community of 345 Million People

As it stands the EC is already an imposing body. Along with the USA and Japan it is considered one of the world's three great economic 'poles'. It is not its physical extent which is the most impressive, though it stretches 2500 kilometres from Copenhagen to Seville and 2900 kilometres from Dublin to Athens. Its area is 2.4 million square kilometres, a quarter the size of the USA and less than 2 per cent of the other continents. It is the Community's population and economic power that attracts attention and respect from around the world. In 1990 there were 345 million people living in the 12 states, many more than in the USA. At the same time annual production was more than 4800 billion dollars in value, a little more than a quarter of the world total (25.4 per cent); equal to the USA and rather more than Japan (15 per cent). For the moment the Community remains politically weak, owing to a lack of real commitment to union, but its demographic and economic power is nevertheless considerable.

### A Group of Advanced Countries

Viewed from a global perspective, the members of the EC are firmly placed in the group of more advanced countries. Their gross national product per capita was just over 14,000 dollars a year in 1990; less than in Japan or the USA, but about four times the world average. At 81 per cent, the level of urbanization was twice the world figure.

The member states are particularly advanced in demographic terms. The demographic transition began very early here. The decline in the death rate began in a rather halting way during the eighteenth century. Similarly, the decline in the birth rate was also precocious, especially so in France where there is evidence of fertility control even before the Revolution of 1789. In the remainder of western Europe the secular decline of fertility began in the last quarter of the nineteenth century, appreciably earlier than in North America, Oceania and Japan. As a result of this evolution, birth and death rates are among the lowest in the world. Life expectancy at birth in 1990 was almost 76 years, two years more than in the developed countries taken as a whole and 12 more than the global figure. Infant mortality has reached a particularly low level: at nine per thousand live births it is half that in the industrialized countries in general. As for the total fertility rate, this shows an average of 1.6 children per woman, fewer than in the developed countries as a group (2.0) and much fewer than the world average (3.5).

This particularly low level of fertility has had a number of consequences. Population growth has been slight over the last 20 years and could soon reach zero. In 1990 the population grew by only 0.2 per cent whereas in the developed countries the growth was 0.5 per cent and in the world as a whole it was 1.8 per cent. Demographic ageing is another direct result. People aged over 65 already represent 14 per cent of the population compared with 6 per cent in the world as a whole.

## Relatively Diverse Populations

As we have seen, the populations making up the EC show certain similarities and seem relatively alike when compared with those of other parts of the world. Nevertheless, they remain quite diverse. European society is the product of a long and warring history where populations from different languages, religions and economic interests have been opposed. Within the Community these differences are still apparent in spite of a process of convergence which is reducing them little by little.

First of all, there are large differences in size and therefore in demographic weight (tables 1.1 and 1.2). From this point of view Germany has become the heavyweight since its reunification, with almost 80 million inhabitants. The three states of France, Italy and the UK have similar weighting with 56–7 million inhabitants each. Spain has about 40 million. The others come far behind: the Netherlands has 15 million, Belgium, Greece and Portugal have roughly 10 million each; Denmark has 5 million and Eire 3.5 million. As for Luxembourg, this country is the featherweight with 0.37 million inhabitants, less than an average English county or even a French département.

Next, there is clear disparity in the intensity and the modes of

**Table 1.1** Key indicators for the countries of the European Community

| *Country* | *Area (thousands of km²)* | *Population, 1990 (millions)* | *GDP, 1988 (billions of dollars)* | *GDP per capita, 1988 (thousands of dollars)* |
|---|---|---|---|---|
| Belgium | 30.5 | 9.9 | 148 | 14.0 |
| Denmark | 43.1 | 5.1 | 108 | 20.9 |
| Eire | 70.3 | 3.5 | 31 | 8.8 |
| France | 551.6 | 56.4 | 945 | 16.9 |
| Germany (FRG) | 248.7 | 63.2 | 1,207 | 19.7 |
| Germany (GDR) | 108.3 | 16.3 | 145 | 8.7 |
| Greece | 131.9 | 10.1 | 53 | 5.3 |
| Italy | 301.3 | 57.7 | 820 | 14.3 |
| Luxembourg | 2.6 | 0.4 | 6 | 16.8 |
| Netherlands | 41.9 | 14.9 | 227 | 15.4 |
| Portugal | 92.4 | 10.4 | 41 | 4.0 |
| Spain | 504.8 | 39.4 | 339 | 8.5 |
| UK | 244.1 | 57.4 | 802 | 14.1 |
| Total | 2,371.5 | 344.7 | 4,872 | 14.1 |

*Sources*: *Atlas Statistique, 1990*; World population data sheet, 1990

**Table 1.2** Weighting of the different countries within the European Community

| *Country* | *Demographic weight (population 1990, % of the total)* | *Economic weight (GDP 1988, % of the total)* |
|---|---|---|
| Germany | 23.1 | 27.8 |
| France | 16.4 | 19.4 |
| Italy | 16.7 | 16.8 |
| UK | 16.7 | 16.5 |
| Spain | 11.4 | 7.0 |
| Netherlands | 4.3 | 4.7 |
| Belgium | 2.9 | 3.0 |
| Denmark | 1.5 | 2.2 |
| Greece | 2.9 | 1.1 |
| Portugal | 3.0 | 0.8 |
| Eire | 1.0 | 0.6 |
| Luxembourg | 0.1 | 0.1 |

human occupation of the land; this is the product of ancient inheritance and has tended to increase since the industrial revolution. The population is exceptionally dense in some countries: Belgium has 326 inhabitants per square kilometre and the Netherlands has 361. In sharp contrast, Eire's population density is just 53. The degree of urban concentration is also very variable. The level of urbanization is 95 per cent in Belgium and 94 per cent in the former Federal Republic of Germany where the population is almost entirely urbanized or suburbanized. On the other hand, Portugal with 30 per cent remains largely rural. Even though these simple comparisons are not straightforward, since the level of urbanization is notoriously difficult to define, there remain real differences in the extent of urbanization between Community countries (see chapter 3).

Finally, there are significant demographic variations. The median age of the population varies between 27 and 38 years and the total fertility rate varies between 1.3 and 2.2 children per woman. As for natural growth, it ranges from about 0.6 per cent per year in Eire to a slightly negative figure in the former Federal Republic of Germany.

In spite of certain common processes which tend to create conformity the EC still displays a considerable amount of demographic diversity. The historical legacy of cultural fragmentation will doubtless ensure that this intriguing geographical kaleidoscope of peoples, social values and environments, and with it Europe's distinctive population geography, will persist for several generations.

2

# Evolution of the Population: a Slow Growth

Philip E. Ogden

That Europe should reach an end to rapid population growth was a foregone conclusion.

Frank W. Notestein et al. (1944, p. 46)

With respect to trends in population growth and fertility, the present simultaneity and uniformity of trends and reversals across Western Europe contrasts sharply with the historical diversity in the timing and pace of demographic transition.

Jean Bourgeois-Pichat (1981, p. 19)

The current preoccupation with slow rates of population increase in the European Community (EC) benefits from being viewed in the context of the twentieth century as a whole. While the total population had increased to more than 340 million by the early 1990s, representing considerable spurts of growth in particular places and at particular times, population stability is by no means new. Fears of population ageing, and even decline, were already being expressed in some countries at the start of the century and the demographic trauma of the First World War for the major states was followed by very low rates of growth in the inter-war period. The demographic renewal which came after 1945 may be seen as something of an exception. Whilst Notestein and his colleagues (1944) were wrong in their immediate assumptions of population decline in the post-war years, their general fears about the slowing of population growth through fertility decline have proved correct in the longer term. The twentieth century has been marked by the gradual retreat from the high fertility regimes of the nineteenth century and a sharp decline in Europe's relative demographic importance in global terms.

This chapter aims to provide a broad picture of change from around 1900 onwards, reviewing the evolution of total rates of growth and their major determinants. Two themes are of particular importance: first, the general tendency towards a 'mature' demographic structure, that is, low fertility, the rapid retreat of mortality and their corollary of ageing, achieved through the process usually described as 'demographic transition'; second, though we may make much of continued geographical differences within and between the 12 countries which span cultures as different as Eire and Greece or Denmark and Spain, this century has been marked by growing similarity in demographic trends, so that the Europe of the early 1990s has become much more homogeneous in terms of demographic behaviour. Certainly the lives of Europeans in the late twentieth century are lived very differently in terms of family relationships, longevity, proneness to disease, mobility and geographical location, and we should not allow the demographic data to obscure our appreciation of these changes in the most fundamental of human experiences. Twentieth-century Europe has been marked by great economic and political upheavals, including the two world wars, and yet a powerful, if quieter, theme running through the century has been this change in demographic behaviour, affected only at the margins by politics and policy. It should also be pointed out at the outset that mobility has undergone important changes during the century. There has been a massive shift from rural to urban residence in all countries. In addition, many countries which had been exporters of people to the New World became countries of immigration, especially in the post-war years. There may be something distinctly 'European' in the nature of demographic change in the longer term, particularly the early decline of marital fertility, but the processes described here are very much part of the wider process of change in the developed world as a whole. Whilst there are interesting differences between the Community and the countries of eastern and central Europe, the 12 states share remarkable similarities, and the demographic 'system' of which they are a part is of much wider geographical extent.

## General Demographic Evolution

The 12 countries that currently make up the EC have remained of very similar relative size during the century: the five largest – France, Germany, Italy, Spain and the UK – accounted for about

87 per cent of the total in 1900 and 84 per cent in 1989, and it is to these countries that we shall give particular attention. All 12, with the exception of Eire, have experienced substantial population growth throughout the century, although, as chapter 1 has indicated, the relative importance of Europe as a whole in global demographic terms has been in steep decline. The Community is a very significant bloc, on a par with the USA and the former USSR, but the impetus of population growth has lain very firmly with the Third World during the last three decades.

Table 2.1 reveals the very different rates of growth for the 12 countries since 1900. There are two broad categories: those countries such as France, Germany, the UK and Belgium where the demographic transition was already well under way by the end of the nineteenth century and growth during this century was more limited; and countries like Italy, Spain, Portugal and Greece in southern Europe and Denmark and the Netherlands in the north where strong growth was still to come. Eire was the perpetual exception since it registered population decline for much of the century. Bourgeois-Pichat's observation which appears at the beginning of the chapter suggests that countries moved towards shared characteristics by different routes. By 1990 there were only relatively small differences in the basic demographic indicators,

**Table 2.1** Populations of the 12 European Community countries, 1900–91 (millions)

| *Country* | *1900* | *1938* | *1960* | *1991*[a] | *1900–91 (% change)* | *1960–91 (% change)* |
|---|---|---|---|---|---|---|
| Belgium | 6.7 | 8.4 | 9.2 | 9.9 | 47.8 | 7.6 |
| Denmark | 2.5 | 3.8 | 4.6 | 5.1 | 104.0 | 10.9 |
| Eire | 3.2 | 2.9 | 2.8 | 3.5 | 9.4 | 25.0 |
| France | 38.5 | 42.0 | 45.7 | 56.7 | 47.3 | 24.1 |
| Germany (with GDR) | 56.3 | 68.6 | 72.7 | 79.5 | 41.2 | 9.4 |
| Greece | 2.4 | 7.1 | 8.3 | 10.1 | 320.8 | 21.7 |
| Italy | 33.6 | 43.6 | 49.6 | 57.7 | 71.7 | 16.3 |
| Luxembourg | 0.23 | 0.30 | 0.31 | 0.38 | 65.2 | 22.6 |
| Netherlands | 5.1 | 8.7 | 11.5 | 15.0 | 194.1 | 30.4 |
| Portugal | 5.4 | 7.7 | 8.9 | 10.4 | 92.6 | 16.9 |
| Spain | 19.1 | 25.3 | 30.3 | 39.0 | 104.2 | 28.7 |
| UK | 38.7 | 47.5 | 52.5 | 57.5 | 48.6 | 9.5 |
| EC of 12 | 211.7 | 265.9 | 296.4 | 344.8 | 62.9 | 16.3 |

[a] Mid-year estimate.

*Sources*: Reinhard et al., 1968, p. 686; Hoffmann-Nowotny and Fux, 1991, p. 73; Rallu and Blum, 1991; *Population et Sociétés* 259 (July–August 1991)

even in traditionally very diverse cultural circumstances. Thus crude birth rates varied only marginally around the European average of 11.8 per thousand, between Eire's 15.1 and Italy's 9.8, while infant mortality varied between 7.1 in the Netherlands and 11.0 in Portugal (table 2.2; see also Nobile, 1990). By 1989, average life expectancy at birth was over 76 years for women in all countries, and over 80 in France and the Netherlands. For men, only the former German Democratic Republic had a value below 70 years and the other 12 countries lay within two years of the 72.7 years average. Moreover, while substantial variations in a number of demographic variables remained both geographically and by social class, the tendency in most of the major measures, taken over the longer term, has been towards homogeneity. These figures compare in an interesting way with those for the turn of the century. Life expectancy at birth was under 50 years for both

**Table 2.2** Population characteristics of the 12 European Community countries in 1989–90

| *Country* | *Crude birth rate, 1990 (per 1,000)* | *Crude death rate, 1990 (per 1,000)* | *Total fertility rate, 1990 (children per woman)* | *Infant mortality rate, 1990 (per 1,000)* | *Life expectancy (male) c.1989* | *Life expectancy (female) c.1989* |
|---|---|---|---|---|---|---|
| Belgium | 12.4 | 10.5 | 1.62 | 8.0 | 72.4 | 79.0 |
| Denmark | 12.4 | 11.9 | 1.67 | 8.4[a] | 72.0 | 77.7 |
| Eire | 15.1 | 9.1 | 2.18 | 8.2 | 71.0 | 77.0 |
| France | 13.5 | 9.4 | 1.80 | 7.2 | 72.4 | 80.6 |
| Germany | 11.3 | 11.3 | 1.46 | 7.5[a] | n.a. | n.a. |
| FRG | 11.5 | 11.2 | 1.48 | 7.5[a] | 72.6 | 79.0 |
| GDR | 10.5 | 12.1 | 1.41 | 7.5[a] | 69.8 | 75.9 |
| Greece | 9.9 | 9.2 | 1.45 | 10.0 | 72.6[b] | 77.6[b] |
| Italy | 9.8 | 9.3 | 1.27 | 8.6 | 73.2 | 79.7 |
| Luxembourg | 12.9 | 10.0 | 1.61 | 9.9[a] | 70.6[b] | 77.9[b] |
| Netherlands | 13.2 | 8.6 | 1.62 | 7.1 | 73.7 | 80.2 |
| Portugal | 11.2 | 9.9 | 1.43 | 11.0 | 70.9 | 77.9 |
| Spain | 10.2 | 8.6 | 1.33 | 7.6 | 73.2[b] | 79.8[b] |
| UK | 13.9 | 11.2 | 1.85 | 7.9 | 72.2 | 77.9 |
| EC of 12 | 11.8 | 10.1 | 1.58[a] | 8.1[a] | 72.7 | 79.1 |
| World | 27.0 | 9.0 | 3.4 | 68.0 | 65.0[c] | |

n.a., not available.
[a] Data for 1989.
[b] *c.*1986.
[c] Male and female.
*Source*: *Population et Sociétés* 261 (October 1991) p. 6

sexes and infant mortality was in excess of 100 per thousand live births (Reinhard et al., 1968, p. 494). Distinctions within and between countries were much sharper, as were contrasts among social groups.

## Population Change Before 1945

### *Demographic Transition*

Recent research has allowed us to be much more specific about the origins of the major trends in European population in the twentieth century. In particular, the Princeton European Fertility Project (Coale and Watkins, 1986; Watkins, 1991) has concentrated on the analysis of fertility decline from the mid-nineteenth century onwards, and thus on assessing the progress of the demographic transition in general. Decline of fertility went hand in hand with, and was frequently preceded by, the decline of mortality (Schofield et al., 1991). Both were complex processes affected not only by the changing nature of demographic behaviour itself but by the wider socio-economic environment as well. High mortality rates affected the younger as well as the older age-groups and were a crucial means by which the move towards the stabilization of population growth was moderated. The variable level of fertility in say the 1870s and its subsequent decline was influenced by a variety of factors. For example, age at marriage had a direct effect on the likelihood of child-bearing and on other factors such as the age structure of the population, the proportions marrying and the number of illegitimate births. These in turn were related to differing cultures and levels of economic development. In addition, the nature of, and access to, means of family limitation – either via sexual abstinence or by the use of appliance methods of birth control – have been recognized as important facilitating aspects of the secular decline of marital fertility. During the demographic transition which began in most regions of western Europe in the last decades of the nineteenth century, the key factors appear to have been the reduction in the number of large families, in the number of illegitimate births and in the frequency of marriage, an increase in average age at marriage and the growth of emigration (Festy, 1979, p. 180). The geography of fertility decline reveals that change occurred not just at the national scale but at the level of regions, where cultural differences were strongly in evidence in Europe at the turn of the century.

The data for European provinces assembled by members of the Princeton European Fertility Project provide a particularly valuable insight into the geography of fertility variation and decline (Coale and Watkins, 1986, especially pp. 31–177 and the maps that appear at the end of the volume). In the case of France, for example, by 1900 almost every province had experienced a very substantial decline in a process that had begun as early as the late eighteenth century (van de Walle, 1974). Though there were still patches of high marital fertility in some areas, especially in the more traditional upland regions of the Massif Central, Brittany or the Alps, the extent of decline in a country still very rural and regionally diverse is quite remarkable, and for this reason France has always been something of a demographic enigma. Nevertheless, by 1900, it was no longer alone: low and apparently controlled marital fertility was already a feature of much of England and several other provinces in southern Belgium, Germany and Italy. A sustained decline in marital fertility had been experienced by at least half of Europe's provinces by 1903. In terms of the 12 Community states under discussion here, there are clear signs of a core and periphery. Areas of later fertility decline tend to lie on the periphery: Eire, Scotland and parts of Wales to the west; parts of Spain, Portugal and southern Italy to the south.

Regional variation within the countries was also of great significance. In the case of Germany, for example, Knodel (1974; see also Woods, 1987) has suggested that this spatial patttern was inversely associated with the level of urbanization and positively linked to the proportion of the population that was catholic. Protestant areas had low fertility levels and so did the largely urban Jewish populations. Yet the rate of fertility decline does not itself correlate well with these factors, or others such as infant mortality and industrialization. For Knodel and van de Walle (1979; also Coale and Watkins, 1986, pp. 390–419), indeed, the Princeton studies show that fertility decline took place under very diverse socio-demographic conditions and there was no simple correlation between development and demographic change. Despite the picture of diversity revealed by the various maps of regional fertility patterns in 1870 or 1900, Knodel and van de Walle (1979) rightly point out that

> the striking factor that the countries of Europe had in common when fertility declined was time itself . . . : with the exception of the forerunner, France and a few stragglers, such as Ireland and Albania, the dates of the decline were remarkably concentrated. The momentous revolution of family limitation began in two-thirds of the province-sized administrative areas of Europe during a thirty-year period, from 1880–1910.

Schofield and Reher (1991, p. 6) have made a similar point in their account of the mortality transition which

> raised life expectancy by more than ten years over a period of only three decades. . . . Even though mortality in wealthier countries was lower and declined at somewhat faster rates, the entire process took place everywhere on the continent during a very short period of time, much as had occurred with fertility.

They also emphasize the reduction in the wide regional differences in mortality that 'had been present for at least two or three centuries'. Thus, the inter-regional range of life expectancy at birth in Spain from 25 years to over 40 years in 1860 or in infant mortality from 290 per thousand births in Stockholm to under 100 per thousand births in certain rural areas were soon to disappear in the twentieth century (Schofield and Reher, 1991, p. 4).

Knodel and van de Walle (1979) also emphasize the importance of cultural factors – such as common dialect and common customs – rather than socio-economic variations in determining the pace of change with respect to, for example, the geographical diffusion of attitudes to fertility and contraceptive practice. They conclude that 'there is greater similarity in fertility trends among provinces within the same region but with different socio-economic characteristics than is true among provinces with similar socio-economic characteristics but located in different regions' (p. 236). Certainly the subsequent evolution during the twentieth century leaves little doubt that fertility decline became a major cultural feature of almost all regions. By 1960 the Princeton fertility indices show that inter-province variation was relatively slight compared with the nineteenth century. Overall fertility was low and some form of family limitation was obviously practised by at least a majority of married couples in all European provinces.

Watkins has argued recently that demographic behaviour shifted in scale from the sub-national, provincial level of diversity to a point at which national boundaries became more important. Thus, the

> reduction in within-country demographic diversity was paralleled by a trio of macro-level changes: the integration of national markets, the expansion of state functions and nation-building. It might seem that these processes would have little to do with such private behaviors as marriage and childbearing. But [they] increasingly drew local communities into national networks.
>
> (Watkins, 1990, p. 262; see also Watkins, 1991)

### *The Impact of the First World War and Inter-war Population Changes*

Contemporaries had often remarked upon the significance of fertility decline for social and political issues even before the First World War (Huss, 1988), but the conflict itself both aggravated the 'problem' and heightened the awareness among politicians and some parts of public opinion about the potential consequences of slowing population growth. The demographic impact of the war was indeed considerable. For Britain, war losses 'touched virtually every household . . . and nearly every family was diminished by the death in combat of a father, a son, a brother, a cousin, or a friend'. The war 'had cut a swathe through an entire generation' (Winter, 1985, p. 305). Losses in continental Europe and overseas were similarly severe. Winter (1985, p. 75) has estimated that 9.45 million died in military action including 2 million Germans, 1.8 million Russians, 1.3 million French and 723,000 British or Irish. To these we must add not only the large number of injured who subsequently died, but also civilian deaths and the births lost as a result of the disruption caused to normal family life by the war. War losses were extremely age-specific: in Britain almost 60 per cent of war deaths were in the age range 20–9 (Winter, 1985, p. 81). In France the population was 'mutilated like a living organism: the age pyramid shows this injury to its flank, like an axe blow' (Reinhard et al., 1968, p. 489). Many countries lost a large proportion of their most economically and demographically active men and this loss had powerful repercussions in the years after 1918.

Immediately after the First World War there was something of a 'baby boom' with at least some of the conceptions prevented by the war taking place. There were also further improvements in mortality in many countries: infant mortality in France, for example, declined from 126 per thousand births in 1914 to 69 per thousand births 20 years later, by which time comparable figures for Britain were 57; for Germany, 68; and for Italy, 101. Life expectancy at birth improved generally: from 48.5 years in 1910 to 54.3 in 1930 for France; from 51.5 to 58.7 for England and Wales; and from 47.4 to 59.9 for Germany (Reinhard et al., 1968, pp. 494, 505). Improvements in nutrition in many countries, whilst unevenly spread by social class, were matched by impressive medical progress in the fight against infectious disease, in surgical techniques and in sanitary and public health measures. However, the baby boom did not last and, despite the decline in mortality, rates of natural population increase soon reached their lowest levels ever.

By the 1930s fertility had fallen to below replacement level in many countries (Teitelbaum and Winter, 1985, pp. 70–1) and the prospect of further decline was likely in those countries where the *classes creuses* – gaps created in the age structure – were most evident, as this effect was mirrored in the next generation. During the mid-1930s France experienced a natural population decrease and growth rates were also very low in Britain. By 1939, of the 90 French départements, 60 were recording an excess of deaths over births. Ageing was a concern in many countries, as was the continued equation of demographic health with military strength. When the threat of war loomed again the French noted with alarm that while they had only 2.6 million in their 20s, the Germans had 6.5 million and the Italians 3.5 million.

Reactions to the demographic crisis of the 1930s fell into two groups. Germany and Italy, on the one hand, under the guidance of National Socialism developed sharply interventionist, racist policies. Both feared population decline, the first through declining births and the second through emigration and high mortality. By 1939 both had seen a rise in their rates of population increase which were a good deal higher than those of their neighbours. In Germany, the birth rate rose from 1933 onwards under the influence of active pro-natalist policies, encouraging marriage and the family and outlawing abortion, all within the well-known selective Nazi ethic, although the overall demographic results were rather precarious (Reinhard et al., 1968, p. 518). In Italy, the Fascists took measures to limit emigration, reduce mortality and encourage marriage and the family, all of which seemed to be having the required effect during the late 1930s but were brought to an abrupt halt a few years later.

Elsewhere in Europe concern over the declining birth rate was strongly expressed, though policy took rather less interventionist forms as governments were slow to react. The climate of the debate in Britain or France is well represented by contemporary propaganda and the titles of books published at the time: for example, L. C. Money's *The Peril of the White Race* (1925), D. V. Glass's *The Struggle for Population* (1936), Enid Charles's *The Menace of Under-Population: A Biological Study of the Decline of Population Growth* (1936) and J. J. Spengler's *France Faces Depopulation* (1938). In both countries there was a veritable flood of publications within, for example, the expanding field of eugenics or on specific issues such as ageing and the means by which fertility might be raised (see Ogden and Huss, 1982; Teitelbaum and Winter, 1985; Huss, 1990; Schneider, 1990; Thane, 1990).

### *The Evolution of Migration*

The inter-war period was also marked by significant shifts in migration both within countries and beyond their borders. Given the surge of immigration to many of the countries of the EC after 1945, and the frequently intense debate that has resulted, the fact that Europe had for long been a mass exporter of people is often forgotten (Baines, 1991). Indeed, most estimates put the number of migrants from Europe as a whole during the period 1800–1930 at some 40 million, a phenomenon of such magnitude as to produce economic and cultural change at the global scale as well as profound social consequences in the countries of origin. A major part of this emigration was, of course, absorbed by the USA: before about 1885, the countries of north-west Europe were the most important suppliers, but after that date southern and eastern Europe saw the departure of millions of their citizens in what came to be known as the 'new migration'. Thus, Italy lost perhaps 9 million people overseas between 1876 and 1925, while Spain and Portugal sent large numbers to Latin America, notably Argentina and Brazil. The rate and timing of departures was linked both to the differential rates of industrialization in the origin areas (Potts, 1990, p. 131) and to conditions in the countries of destination.

As in other respects, the First World War proved a turning point. It marked the beginning of the end of mass migration from Europe, and a subsequently lower level of migration involved more movement within Europe than beyond. France, for example, became a major importer of labour from the neighbouring countries of Italy, Belgium, Spain, Poland and Switzerland (Ogden, 1989), building on nineteenth-century origins. While the steep decline in emigration marked by the First World War itself was temporarily reversed after the end of hostilities, emigration never regained its former momentum: some 40 per cent of population growth in the USA had been accounted for by immigration during the period 1880–1910, but by 1910–20 the figure fell to 17 per cent (Reinhard et al., 1968, p. 415). Among the causes of this change were the slowing of population increase in Europe, the improvement in living standards and the imposition of immigration controls in receiving countries, especially the USA. The economic crisis of the 1930s further undermined migration flows, while some of the new political regimes in Europe, in Italy for example, launched vigorous campaigns to discourage emigration and attract return migrants. Thus, a regulating mechanism of great social and cultural significance before 1914 was transformed during the inter-war years and set the scene, as we shall see below, for a new era of mass migration after 1945.

## Population Change After 1945

The Second World War was traumatic and destructive in so many well-known ways for much of continental Europe and had massive demographic consequences in terms of lives lost, population redistribution and labour shortage. The post-war years, however, were to contain a number of surprises which confounded much of the pre-1939 pessimism about demographic trends. Though the general trend that we can observe during the period since 1945 as a whole is a remarkable move to homogeneity and synchroneity in demographic responses in almost all the 12 countries, a number of very distinct features did persist at the national and regional levels. Overall, we must recognize the immense importance of the reduction of both fertility and mortality to historically unprecedented low levels, the rise of divorce and the liberalization of contraception and abortion in most countries, the increasing separation of sexual activity and reproduction and the waxing and waning of both international and rural to urban migration. By the 1990s the Europeans had become one of the healthiest, longest-living, most urbanized populations in the world exercising an unprecedented degree of control over their demographic fortunes. Yet social and geographical patterns still have much to teach us and there are interesting recent trends in a number of demographic variables. Most of these themes are dealt with in more detail in the following chapters; here I shall mention only the most significant developments.

### *Overall Population Growth*

Given the generally pessimistic tone of the debate on fertility in the 1930s and the ready assumption in many countries that a new and largely irreversible era of the demographic transition had been reached, one with low rates of natural increase and a rapidly ageing population, there was some surprise that in the years 1945 to the later 1960s the population of the 12 Community countries in fact grew substantially. This was caused by several factors: an increase in marriage and in fertility associated not just with a post-war baby boom but with a more sustained trend of increasing family size; a rapid increase in immigration to several north-west European countries, often those with the lowest pre-war rates of population increase; and a sustained decline in mortality. However, since the mid-1960s growth rates have been dramatically curtailed. Although mortality has continued to fall everywhere, fertility decline has become the key dynamic feature, accentuated in some countries by a decline in immigration. This has been described by van de Kaa (1987) as Europe's second demographic transition, associated as it has been with marked changes in individual attitudes and behaviour. Fertility series since 1945 divide into two quite distinct periods, approximately before and after 1965. Before that date there was still considerable variation; after 1965 the uniformity of the trend was very marked, suggesting the operation of highly generalized cultural and behavioural processes. Fertility decline has affected all the countries of the Community. By 1988 the number of births stood at 3.87 million per year compared with 5.6 million in 1964. (In 1988 there were 3.6 million births in the USA, 5.2 million in the USSR, 22.5 million in the People's Republic of China and 26 million in India.)

We may usefully draw some comparisons between two dates, 1960 and 1987–8, to show the extent of recent changes in the rates of population growth and their principal components. For the 12 Community countries as a whole the rate of increase declined from 8.2 per thousand in 1960 to 3.5 per thousand in 1988 (figure 2.1(a)) and their share of the world's population declined from 9 per cent to 6 per cent. Variability among the countries was, and remains, considerable (figure 2.1(b)). In 1960, while Portugal and Eire were experiencing considerable population losses (−12 and −5 per thousand respectively), at the other end of the scale most were gaining strongly: West Germany at 12 per thousand and France at 9.6 per thousand. By 1988, the range of growth rates was still considerable, between Eire's −5.2 per thousand and Luxembourg's 8.6. Trends in mortality and fertility have been very similar and much of the current variability derives from sharp fluctuations in migration rates. There exists in fact a number of quite distinct patterns of change over recent decades. Figure 2.2 selects a sample of countries to illustrate the processes at work. Thus, countries such as Portugal and Greece joined in Italy's and Spain's experience of seeing emigration curtailed but lower rates of growth produced by a declining natural increase. It is remarkable that by the late 1980s countries like Greece and Italy had the lowest rates of natural increase in the Community. Eire is and remains very much a special case, with high but declining rates of natural increase and emigration tailing off in the 1970s but re-emerging strongly by the late 1980s. For West Germany, an exceptionally steep decline in natural increase and the curtailing of immigration brought about an actual population decline for several years in the 1970s and 1980s. Finally, population growth continued at a relatively high level, aided by immigration during the 1960s and early 1970s and by a less than average decline in natural increase. The various parts of figure 2.1 also indicate that the most recent years, 1985–8, hint at new trends in population growth: for example, a revival in natural

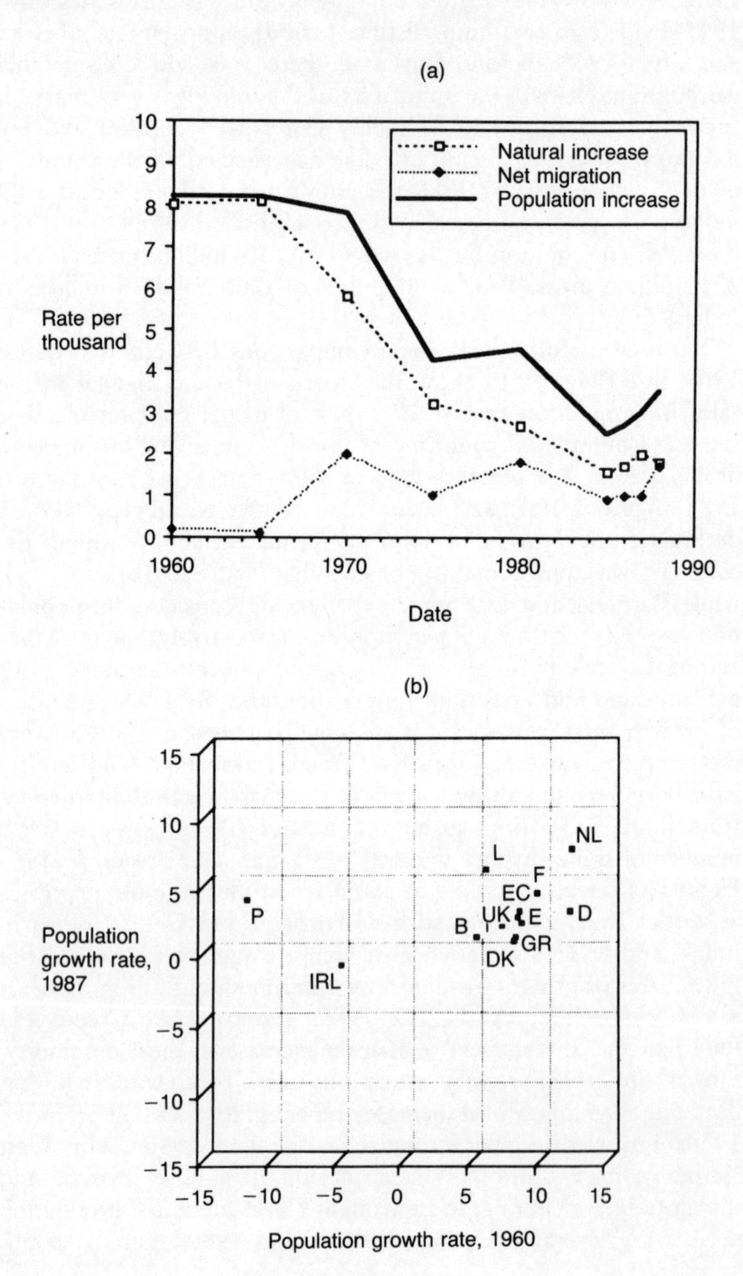


**Figure 2.1** (a) The components of population change in the European Community since 1960; (b) the relationship between rates of population growth (per thousand) in 1987 and 1960.

growth in the UK, in immigration to West Germany and in emigration from Eire (see Hoffmann-Nowotny and Fux, 1991).

Although there are important differences in post-war experience of demographic change as revealed by figure 2.2, the extent of convergence by the 1980s remains the most significant feature and is the theme that flows through many of the subsequent chapters of this book. Where differences persist (for example in fertility, see chapter 5), the trend is generally towards convergence, although new patterns of behaviour with respect to divorce, illegitimacy or cohabitation are by no means universally established. Convergence is well illustrated by the evolution of infant mortality and life expectancy at birth since the Second World War (see chapter 4, especially figures 4.1 and 4.2). In the first case, reduction has been sharp and sustained such that the national level variations among countries are extremely small. In the second case, though overall levels of life expectancy still vary substantially, the trend has been similar and the differences are now less marked than in 1950. This theme of convergence extends beyond the 12 EC countries, but the countries of eastern Europe have performed significantly less well than their western neighbours, especially in terms of life expectancy at birth.

### *Migration*

A further significant change in the demographic growth of countries of the present EC is the role of migration which, with low rates of growth overall, may assume particular significance in the demography of some countries. From 1945 to the early 1960s, the countries could be divided into two camps. On the one hand were those whose economic growth depended upon a labour force recruited abroad, especially France, Germany, the UK, and some of the smaller countries of north-west Europe. On the other hand, the countries of southern Europe became major exporters of labour, especially Italy, Spain and Portugal. Eire continued also its long tradition of emigration. The importing countries drew their labour not only from their European neighbours but also from beyond Europe, usually in relation to their colonial or ex-colonial links. For example, France drew from North Africa; the UK from the Caribbean and South Asia. West Germany, in contrast, at first absorbed a large number of migrants from the east after the division of Europe and subsequently turned to Turkey, Yugoslavia and Greece. By this process of migration, the countries of north-west Europe saw the establishment of a large ethnic minority population in many countries, concentrated especially in the major cities and above all the capital cities.

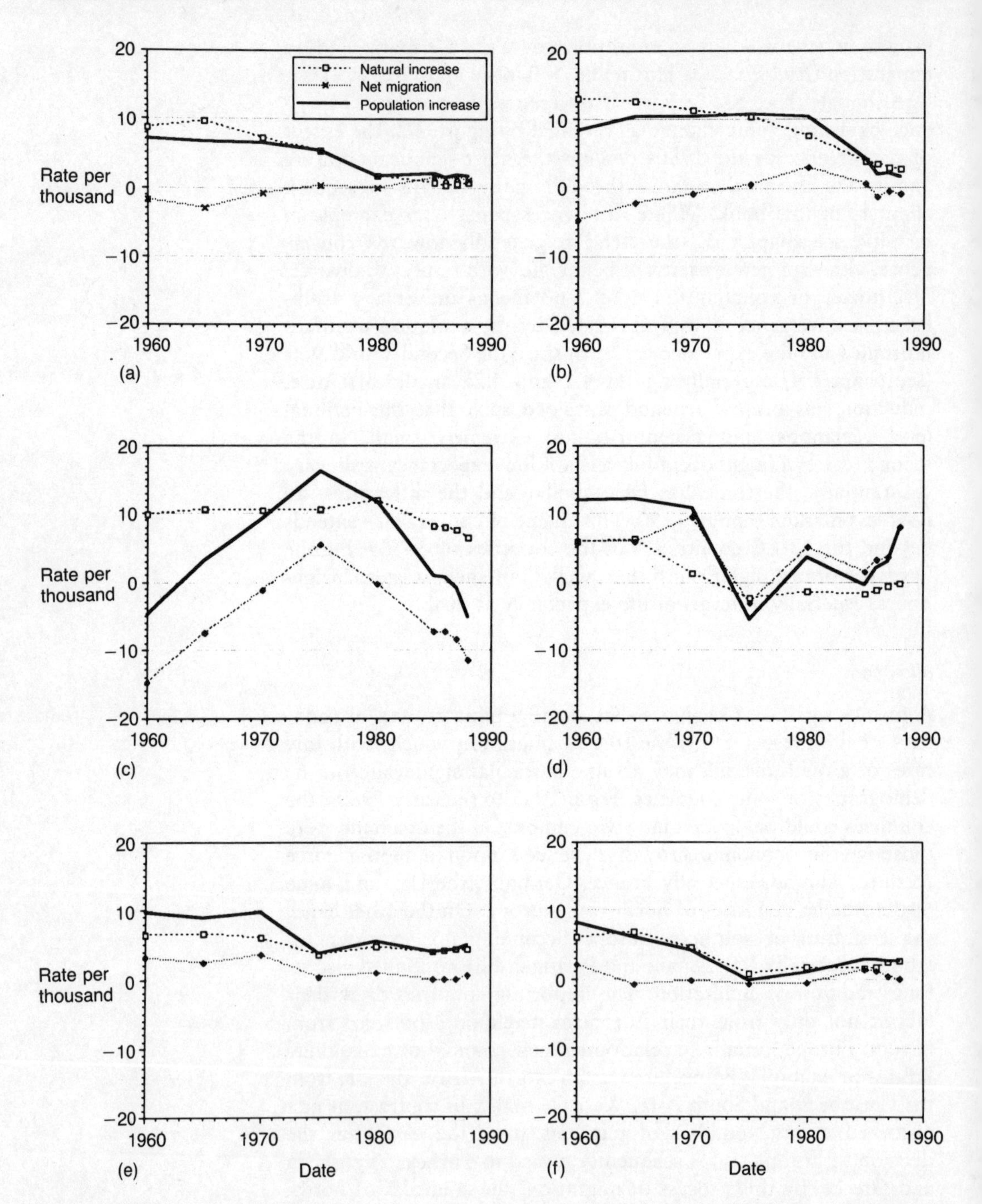


**Figure 2.2** The components of population change since 1960: (a) Italy; (b) Spain; (c) Eire; (d) Germany (FRG); (e) France; (f) UK.

Three factors have intervened to alter this pattern of migration during the last two decades. First, the economic downturn of the early 1970s, associated with the oil crisis, is usually credited with being the cause of a significant decline in labour immigration, as many countries tightened their immigration policies. Second, and perhaps of more importance, there were changes of a fundamental kind in labour needs, especially the decline of manufacturing and the rise of service industries. This has meant an increasingly feminized immigrant labour force and a relatively greater role for skilled migrants (Cohen, 1987; OECD, 1987a). Third, the collapse of the major political divide between eastern and western Europe has ushered in a new period of uncertainty in migration pressures. Thus, an existing demographic pressure from the south has been joined by political and economic pressure from the east. The free market of 1993 has brought the question of immigration control within and from beyond the Community into sharp focus as a policy issue for the member states (van de Kaa, 1991). One of the consequences of these three factors has been a dramatic change in the geography of migration in the Europe of the early 1990s. Many of those countries that had been exporters of labour began to experience immigration, for example Italy and Spain, not just of their own citizens returning home but also of new migrants from the Third World (Uner, 1991). The changing contribution of migration is shown clearly in figure 2.2.

### Conclusions

The history of Europe's population in the twentieth century is marked, above all, by the mastery that has gradually been established over both mortality and fertility. Contemporary trends benefit from being placed in this longer-term perspective. While it would be misleading to depict either the past or the present in terms of geographical uniformity, as this chapter has demonstrated, the long-term trend in the present century towards demographic convergence is well established. The distinctive demographic regime of the late nineteenth century spreads, of course, much wider than the present boundaries of the EC. Variations within countries have been much reduced but there remain important differences from country to country, related, for example, to national welfare and housing policies (D. A. Coleman, 1991). The dry demographic statistics reveal fundamental changes in the attitude of Europeans during the twentieth century towards children and the family, and in their experience of death, disease and ageing. The demographic revolution which began a century or so ago is just as fundamental to understanding contemporary

European society as are recent changes in modes of economic production and urbanization. Finally, however, it should be emphasized that demographic behaviour in the 1990s continues to evolve with important changes in marriage, divorce, illegitimacy, disease patterns and migration making themselves apparent in diverse ways in the regions of the EC.

# 3

# Geographical Distribution and Urbanization

A. G. Champion

The European Community's (EC's) population is by no means evenly distributed, either between countries or between the various parts of the 12 individual member states. Moreover, the geographical patterns changed considerably during the past half-century and seem likely to continue to evolve rapidly through the 1990s and beyond. A central feature of past population shifts concerns the increase in levels of urbanization and the associated trends in rural population. The theme of spatial polarization has also been raised on the broader scale of core–periphery relationships, notably between regions within countries but increasingly across the regional map of the whole Community.

This chapter begins by demonstrating the very uneven distribution of population within the Community and goes on to consider the way in which this can be related to patterns of urbanization and rural population change. It then turns to an examination of metropolitan trends and counter-urbanization. Finally, the main features of the evolving map of population change are described and alternative interpretations are discussed. Where possible, the statistical data refer to the EC as constituted in 1991, including the former German Democratic Republic within the united Germany.

The underlying purpose is to provide a basis for anticipating future developments at the urban and regional level across Europe. Unfortunately, this is by no means an easy task, given the immense variety of distinctive localities in Europe and the as yet limited availability of up-to-date analyses drawing on the detailed results of the 1990–1 census round. There still remains a great deal of uncertainty over the long-term significance of the trends that became apparent in the 1980s. The matter is complicated by recent developments in mass migration arising principally from the emergence of new east–west and north–south relations. It is clear that during the next few years there will be some very important changes in the geography of Europe's population, changes which will have significant consequences for policy formulation.

### The Dimensions of Population Distribution

The 12 member states of the EC vary considerably in terms of their overal population sizes, ranging from barely 380,000 people in Luxembourg to nearly 80 million in Germany (table 3.1). There are, of course, substantial differences in land area, varying from 2586 square kilometres in Luxembourg to over 500,000 square kilometres in France and Spain. These figures alone are sufficient to demonstrate that, even at the national level, population is by no

**Table 3.1** Size, population density and range of regional densities for European Community countries, 1990

| *Country* | *Area (km²)* | *Population (1,000s)* | *Density (persons per km²)* | *Range of densities for NUTS 2 regions* | |
|---|---|---|---|---|---|
| | | | | *Highest* | *Lowest* |
| Belgium | 30,518 | 9,948 | 326.0 | 5,975, Brussels | 52, Luxembourg |
| Denmark | 43,080 | 5,135 | 119.2 | – | – |
| Eire | 68,895 | 3,499 | 50.8 | – | – |
| France | 543,965 | 56,612 | 104.1 | 887, Ile de France | 29, Corse |
| Germany | 356,945 | 79,070 | 222.0 | 3,862, Berlin | 82, Mecklenburg-Verpommern |
| Greece | 131,957 | 10,046 | 76.1 | 999, Attiki | 27, Dytiki Makedonia |
| Italy | 301,277 | 57,576 | 191.1 | 427, Campania | 35, Valle d'Aosta |
| Luxembourg | 2,586 | 378 | 146.3 | – | – |
| Netherlands | 41,863 | 14,891 | 355.7 | 959, Zuid-Holland | 100, Flevoland |
| Portugal | 91,971 | 10,337 | 112.4 | 290, Lisboa e Vale do Tejo | 21, Alentejo |
| Spain | 504,790 | 38,924 | 77.1 | 609, Madrid | 22, Castilla–La Mancha |
| UK | 244,111 | 57,297 | 234.7 | 4,383, London | 9, Highlands and Islands |
| EC of 12 | 2,361,957 | 343,681 | 145.5 | – | – |

Denmark, Eire and Luxembourg are deemed single regions at NUTS 2.
*Source*: EUROSTAT, *Rapid Reports – Regions, 1991/1*

means uniformly distributed around the Community. Indeed, as table 3.1 makes clear, the population density of the Netherlands is seven times that of the least densely inhabited country, Eire.

However, these international differences in population density pale into insignificance beside the extent of regional variations within countries. Given the clustered nature of human settlement, the range of densities will be closely related to the scale of spatial reporting units, but even the relatively coarse grain of the NUTS 2 (Nomenclature des Unités Territoriales Statistiques) regions produces some very marked contrasts, as table 3.1 shows. In the UK, for instance, Greater London's population density of 4383 persons per square kilometre contrasts starkly with the Highlands and Islands Region of Scotland's population density of only nine persons per square kilometre. In Belgium, the range is from 5975 for Brussels to 52 for the Luxembourg province, and in Germany from 3862 for Berlin to 82 for the *land* of Mecklenburg-Verpommern.

The effects of international and inter-regional variations combine to produce the regional map of population distribution. Figure 3.1 shows the population density for the EC in 1988, drawing on data from a hybrid of NUTS 2 and 3 regions according to availability. The map is dominated by certain very high concentrations of population. The English 'megalopolis' stretches from Lancashire and Yorkshire in the north to London and the south coast, and there is an even more extensive zone from northern France to Hanover, with particularly high desities in the Brussels, Randstad and Ruhr agglomerations. Smaller high-density clusters occur in central/southern Germany and southern Italy. These are mainly regions with large cities but not always so, as in the cases of Sicily and Puglia in southern Italy.

Elsewhere, the regional map is punctuated by a few large or very large population concentrations set within larger areas with below average densities. In particular, Paris tends to dominate the rest of France in this way, as Copenhagen does Denmark. In the Nordic countries and eastern Europe the same phenomenon is in operation. If anything, the urban–rural contrasts are even starker in the case of Eire, with its very low densities apart from Dublin. In Iberia, too, Madrid is surrounded by very low densities in Spain's non-coastal *communidades autonomas* and the high densities of the Lisbon and Oporto areas are in sharp contrast with the east of Portugal. Rural Greece and Scotland are also characterized by low densities, but in both of these cases the true extent of urban–rural differences is obscured by the extensive nature of the geographical units which contain Athens and Glasgow, respectively (figure 3.1).

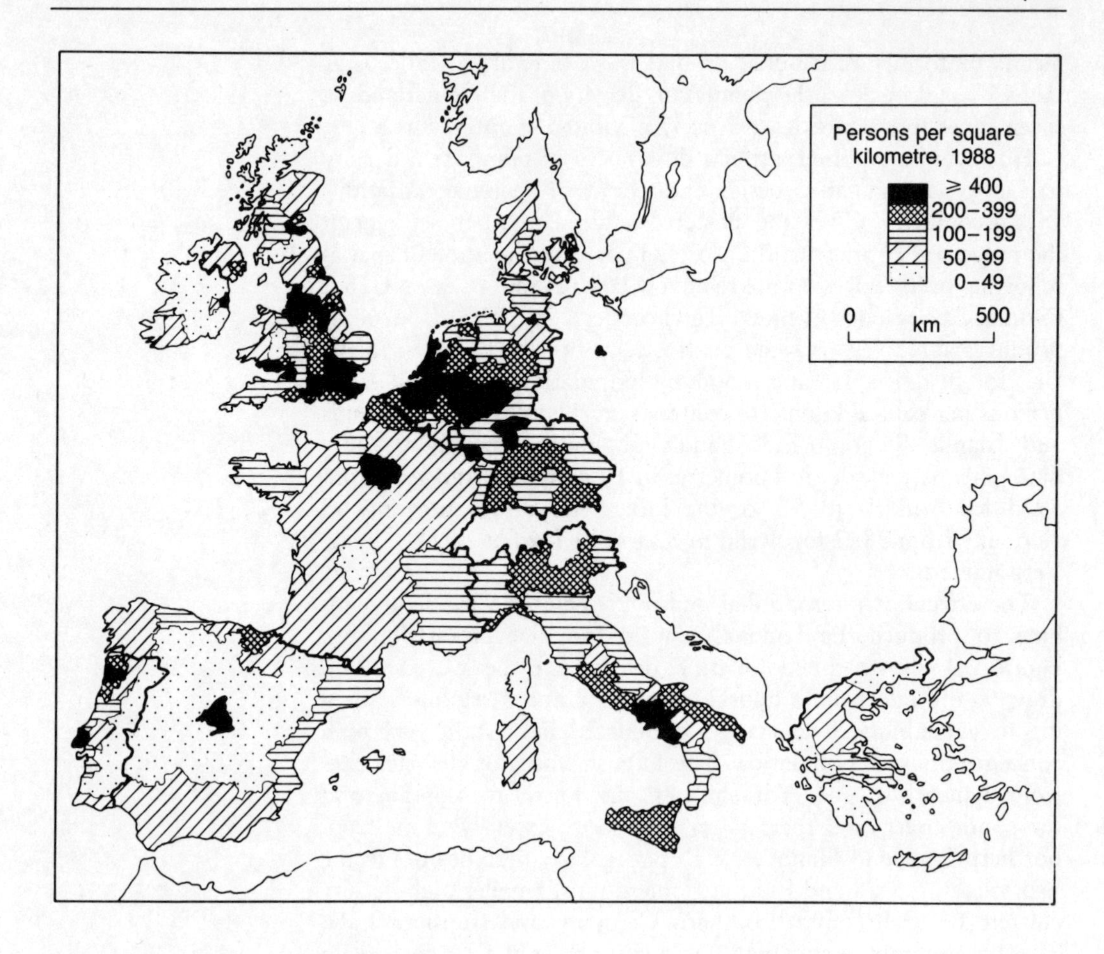


**Figure 3.1** Regional variations in population density, 1988.

## Urbanization and Rural Population Change

Since the eighteenth century and the acceleration of Europe's distinctive industrial revolution, population concentration has become even more closely associated with the process of urbanization, as the proportion of each nation's population living in 'urban places' has grown (de Vries, 1984). Four out of five of the Community's inhabitants now live in 'urban places', variously defined, a significant increase over the level of urbanization that existed 40 years ago. By contrast the rural population has not only shrunk in terms of its share of Europe's population, it has also been declining in absolute terms. Moreover, at the level of this crude urban–rural distinction, it is to be anticipated that these trends will continue, though the rise in urbanization levels will slow down as the latter approach the 100 per cent mark and rates of urban population growth will diminish in line with lower rates of total population growth.

Before setting out fuller details of the European and national experiences of urban growth, it is important to point out that both the measurement and the monitoring of the urbanization process are fraught with difficulties. The principal difficulty concerning measurement is the definition of 'urban place'. The critical definitional issue relates to the minimum size or density that is considered 'urban' and whether the extent of individual settlements is delineated on a purely administrative basis or through some physical or socio-economic criteria. There has as yet been no harmonization of national approaches among the member states. For example, Denmark recognizes an agglomeration of 200 people as urban, while Greece, Italy and Spain apply a 10,000-person threshold to agglomeration, administrative area or the largest settlement in a municipality or commune (United Nations, 1989). The task of monitoring trends depends on the availability of the type of detailed local statistics that are usually found only in population censuses, which makes their availability dependent on each census round.

Table 3.2 shows the proportion of each EC state's population classified as 'urban' according to the definitions adopted by the respective countries, with the 1990 and 2010 projections provided by the United Nations and based on the situation pertaining in each country in the early 1980s. The aggregate statistics for the EC as a whole present a clear picture of substantial and continuing urbanization. It is estimated that in 1990 almost 79 per cent of the Community's population was living in 'urban places'. The proportion can be seen to have grown substantially from its 65 per cent level in 1950, with the fastest increase occurring in the period up to 1971. The United Nations' projections anticipate a further increase of around 5 per cent by 2010 (table 3.2).

At the level of the individual member states, Europe can be divided into two broad groups: those that experienced early industrialization and associated urban growth and those for whom rapid urbanization is a relatively recent phenomenon. The most obvious examples of the former are Belgium, the Netherlands and the UK, with urbanization levels of at least 80 per cent in 1950. Portugal, Greece and Eire are typical of the latter group. Although the process of urbanization has applied throughout the EC, there have been substantial regional and local variations. There has been consolidation of the highly urbanized state of the first group of countries and a rapid shift towards greater urbanization in the latter, most notably in Spain and Greece. The

**Table 3.2** Proportion of population resident in urban places, 1950–2010, for European Community countries (per cent)

| *Country* | *1950* | *1970* | *1990* | *2010* |
|---|---|---|---|---|
| Belgium | 91.5 | 94.3 | 96.9 | 98.1 |
| Denmark | 68.0 | 79.7 | 86.4 | 90.4 |
| Eire | 41.1 | 51.7 | 59.1 | 69.3 |
| France | 56.2 | 71.0 | 74.1 | 79.9 |
| Germany | 71.9 | 78.7 | 84.6 | 87.3 |
| Greece | 37.3 | 52.5 | 62.6 | 72.8 |
| Italy | 54.3 | 64.3 | 68.6 | 76.2 |
| Luxembourg | 59.1 | 67.8 | 84.3 | 89.2 |
| Netherlands | 82.7 | 86.1 | 88.5 | 89.0 |
| Portugal | 19.3 | 26.2 | 33.3 | 46.7 |
| Spain | 51.9 | 66.0 | 78.4 | 85.5 |
| UK | 84.2 | 88.5 | 92.5 | 94.7 |
| EC of 12 | 64.8 | 74.0 | 78.9 | 83.7 |

Figures for 1990 and 2010 are projections based on early 1980s data. Germany and EC of 12 include the former German Democratic Republic. Definitions of urban places vary between the individual countries according to their respective practices.
*Sources*: United Nations, 1989; author's calculations for Germany and EC of 12

projections suggest that the rate of urbanization will proceed most quickly in Eire, Greece and, most notably, Portugal (figure 3.2), but even by 2010 Portugal will still have less than half of its population urban according to its own definition.

The rates of urban and rural population growth which are responsible for these trends in urbanization are shown in table 3.3. These figures reflect overall rates of population growth in the various countries as well as the shifts between rural and urban places. For the EC as a whole, including the former German Democratic Republic, the urban population grew by over 30 per cent (56.7 million) between 1950 and 1970, but by less than half this rate in 1970–90 (13.5 per cent or 32 million). Over the 20 years up to 2010 the level of growth is anticipated to fall again to 8 per cent for the full period, an addition of around 21 million. By contrast, rural areas experienced population drops from 98 million to 72 million over the 40-year period to 1990, with a very similar reduction of around 14 per cent in each 20 year sub-period, and it is anticipated that they will experience an accelerated rate of depopulation during the next 20 years.

Rates of urban growth in individual countries have tended to be highest in those places where the levels of urbanization were

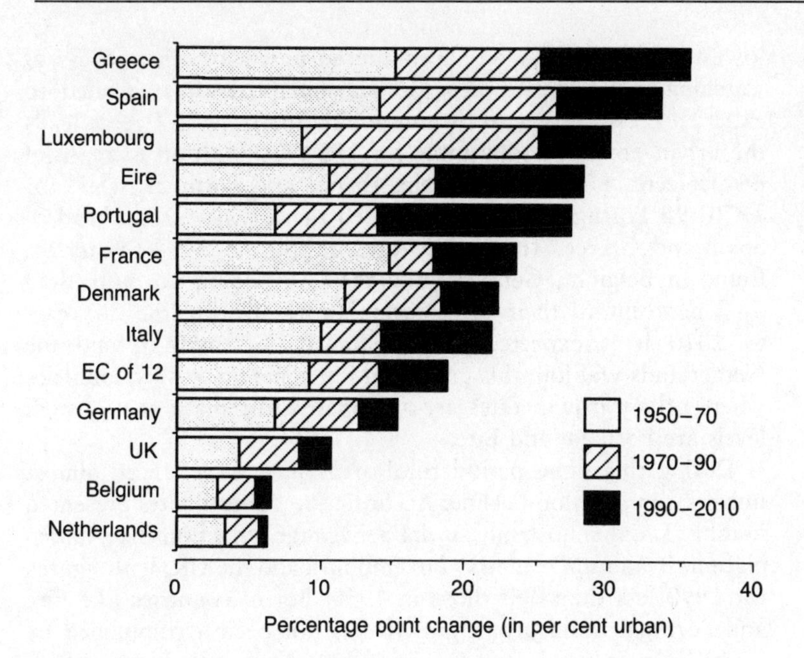


**Figure 3.2** Change in the percentage of European Community countries' populations living in urban places, 1950–2010.

**Table 3.3** Change in size of urban and rural populations, 1950–2010, in European Community countries (per cent)

| | *Urban population* | | | *Rural population* | | |
|---|---|---|---|---|---|---|
| *Country* | *1950–70* | *1970–90* | *1990–2010* | *1950–70* | *1970–90* | *1990–2010* |
| Belgium | 15.2 | 5.8 | 2.3 | −25.6 | −44.6 | −38.5 |
| Denmark | 35.3 | 12.5 | 4.6 | −26.8 | −30.2 | −29.5 |
| Eire | 25.3 | 43.8 | 40.8 | −18.5 | 6.8 | −10.1 |
| France | 53.5 | 15.4 | 14.1 | −19.8 | −1.0 | −17.9 |
| Germany | 26.0 | 5.4 | −0.4 | −17.8 | −24.5 | −20.3 |
| Greece | 63.7 | 36.3 | 18.6 | −12.0 | −10.1 | −25.9 |
| Italy | 35.2 | 13.7 | 11.0 | −10.6 | −6.4 | −24.2 |
| Luxembourg | 31.4 | 34.8 | 4.5 | −10.0 | −46.8 | −32.8 |
| Netherlands | 34.2 | 16.3 | 4.5 | 3.4 | −6.1 | −0.8 |
| Portugal | 46.5 | 44.8 | 47.2 | −1.7 | 2.8 | −16.0 |
| Spain | 53.6 | 38.2 | 15.9 | −14.9 | −26.0 | −28.4 |
| UK | 15.5 | 7.0 | 3.5 | −20.0 | −33.6 | −28.1 |
| EC of 12 | 31.5 | 13.5 | 8.0 | −14.8 | −13.5 | −21.6 |

Rates refer to the full 20-year periods. See also notes for table 3.2.
*Source*: Calculated from United Nations, 1989

lowest in 1950. This has been caused partly by the process of 'catching-up', but also because such countries have tended to experience faster rates of overall population growth. Greece leads the urban growth league table for 1950–70 with an increase of 64 per cent, followed by Spain, France and Portugal, while for 1970–90 Portugal and Eire recorded the highest rates, ahead of Spain and Greece. The lowest rates during the last 40 years are found in Belgium, Germany and the UK, adding no more than 5–7 per cent to their urban populations. During the 20 years to 2010 it is expected that Denmark, Luxembourg and the Netherlands will join this group, and in general the only countries where urban growth rates are anticipated to remain at very high levels are Portugal and Eire.

During this same period rural areas have experienced almost universal population decline. According to the estimates presented in table 3.3, the most substantial percentage reductions have taken place in Belgium, Denmark, Luxembourg and the UK, with figures for 1990 less than half those in 1950. But in countries like Eire and Portugal rapid urban growth has not been accompanied by major losses in rural areas because of the persistently high levels of natural increase among the rural population. These two countries, together with the Netherlands, are expected to record the lowest rates of rural population loss during the next 20 years, but most countries are projected to experience a decline by around a quarter.

### Metropolitan Trends and Counter-urbanization

The process of urbanization, at least in its advanced phase, is often associated with increased concentration in the larger urban places, those metropolitan centres at the top of the urban hierarchy (Berry, 1976). This process has normally reinforced the trend towards even greater spatial polarization of population. Unlike the apparently continuing tendency for the urban proportion of the population to rise, this other aspect of urbanization no longer appears to be as general as it once was. In the first place, transport improvements and other related factors have encouraged suburbanization and the growth of dormitory towns; the geographical extent of individual urban places has grown enormously, so that it is now more usual to speak of functionally defined 'metropolitan areas' or 'functional urban regions' than of physically defined urban places (see, for instance, Hall and Hay, 1980). More recently, there have been indications that, even where urban places are distinguished on the basis of these more general definitions, larger metropolitan areas have been losing population, at least in relative terms, to smaller urban regions in a process that has been termed 'counter-urbanization' (Berry, 1976; Fielding, 1982; Champion, 1989a).

The importance of Europe's larger cities has been demonstrated by a number of studies. According to Cheshire and Hay (1989), for example, in 1981 one in three of the EC's total population were to be found in the 'millionaire cities', that is, functional urban regions with at least 1 million inhabitants, while approximately 70 per cent of the EC population were living in those 229 functional urban regions with populations of 300,000 or more. Estimates provided by the World Bank (1990) indicate the relatively high degree of concentration within Europe's urban population in 1980, with 43 per cent resident in cities with over 500,000 inhabitants and with the largest city in each country accounting for one in six of all urban dwellers on average across the Community (Kunzmann and Wegener, 1991).

The role of the large cities in accommodating Europe's urban population, however, does appear to be in rather general decline according to the World Bank (1990) data for 1960–80 shown in table 3.4. In seven of the 11 countries with cities of over 500,000 people, there was a fall in the proportion of the urban population

**Table 3.4** Degree of urban concentration, 1960–80, in the European Community countries

| | *Percentage of urban population in* | | | | | |
|---|---|---|---|---|---|---|
| | *cities of over 500,000* | | | *largest city* | | |
| *Country* | *1960* | *1980* | *Change* | *1960* | *1980* | *Change* |
| Belgium | 28 | 24 | −4 | 17 | 14 | −3 |
| Denmark | 40 | 32 | −8 | 40 | 32 | −8 |
| Eire | 51 | 48 | −3 | 51 | 48 | −3 |
| France | 34 | 34 | 0 | 25 | 23 | −2 |
| Germany (FRG) | 48 | 45 | −3 | 20 | 18 | −2 |
| Greece | 51 | 70 | +19 | 51 | 57 | +6 |
| Italy | 46 | 52 | +6 | 13 | 17 | +4 |
| Luxembourg | 0 | 0 | 0 | 36 | 33 | −3 |
| Netherlands | 27 | 24 | −3 | 9 | 9 | 0 |
| Portugal | 47 | 44 | −3 | 47 | 44 | −3 |
| Spain | 37 | 44 | +7 | 13 | 17 | +4 |
| UK | 61 | 55 | −6 | 24 | 20 | −4 |

*Source*: World Bank, 1990

accounted for by these cities over the 25-year period, while in France there was no significant change. Only in three were there increases in the share of these large cities: in Italy, Spain and, most notably, Greece – all countries experiencing strong metropolitan expansion as well as rapid urbanization during this period. These changes have been strongly influenced by the share accounted for by the largest city in each country which reveals a rather similar national pattern.

The phenomenon of shifting population down the urban–metropolitan hierarchy has now been documented in some detail for several European countries. Champion (1989b) has shown that when Britain's 280 local labour market areas are grouped by population size there is a very regular relationship between size and 1971–81 growth rate, with the six largest cities losing population fastest and the 52 rural areas gaining the most rapidly. Using French urban agglomerations and rural communes grouped into six size categories, Fielding (1986) has observed the way in which a largely positive statistical relationship between size and net migration rate in 1954–62 was transformed progressively into a largely negative one in 1975–82, when the only agglomerations that were still growing through migration in aggregate terms were those of below 20,000 people. A third example drawn from the West German experience has shown that correlating the rate of net internal migration with population size for the 58 functional urban regions reveals the emergence of a clear counter-urbanization relationship during the latter half of the 1970s (Kontuly and Vogelsang, 1989).

In the early 1980s it was anticipated that this shift in emphasis in population distribution away from the larger metropolitan concentrations towards medium-sized and small settlements would accelerate through the remainder of the decade (see, for instance, Fielding, 1982, p. 2, figure 2). However, this does not seem to have happened on a large scale and indeed there is some evidence of a revival of population growth in some of Europe's largest cities, paralleling the experience of the USA (see Champion, 1992). For example, London's rate of population loss declined markedly in the mid-1980s principally as a result of slower internal migration losses and a substantial increase in net immigration from overseas (Champion and Congdon, 1988). According to the results of the latest French population census, the Paris agglomeration matched the national rate of population growth between 1982 and 1990, a stronger performance than any other group of towns except those with less than 10,000 inhabitants (P. Jones, 1991).

Similar temporal trends have been observed in other studies of the turnaround of metropolitan populations. Most of this work has been designed to monitor migration balances between core and periphery regions in individual countries. It was the work of Vining and Kontuly (1978) that first demonstrated the widespread nature of the swing of net migration away from the major metropolitan regions of the developed Western economies and thereby stimulated the international search for more detailed evidence on counter-urbanization (Champion, 1989a, ch. 1). However, the most recent work in this genre indicates a further reversal, although current patterns are not characterized by quite the same degree of uniformity as those of the early 1970s. According to Cochrane and Vining (1988), even before the end of the 1970s the shift towards the faster growth of more peripheral and rural regions had begun to slow down in much of Europe, especially in the most peripheral countries such as those of Scandinavia. It was also shown that much of southern Europe was still experiencing strong growth in its main metropolitan regions, but during the early 1980s some countries in north-west Europe were still recording net shifts towards their less heavily urbanized regions, most notably in southern Germany and south-eastern France.

### The Evolving Map of Population Change

Recent trends in European urbanization display certain confusing and even contradictory signs which make it difficult to anticipate the way in which regional patterns of population distribution are likely to evolve in the near future. It is expected that more and more Europeans will be living in urban places and that an ever larger proportion of the population will be living in functional urban regions, that is, significant urban centres together with their dormitory towns and the rural areas which make up their commuting hinterlands. What seems far less clear is the size of urban places in these functional urban regions that will experience the fastest gains or losses and also their location. Will they lie in the traditional economic core regions or in the more peripheral areas which appeared to seize the advantage in the 1970s?

A major challenge in the 1990s will be the need to make sense of the regional and urban population trends of the past decade. Figure 3.3 gives some indication of patterns of population change during the 1980s using the same hybrid of regions as figure 3.1. Examination of the patterns within individual countries confirms the mixture of conflicting messages which recent contributions to the literature on metropolitan change and counter-urbanization have been highlighting. In several countries the faster growth rates are likely to be found in their southern regions, but the particular circumstances differ. For example, in Spain and Italy this north–south division is associated with a stronger natural increase rather than higher rates of net in-migration, whereas migration is the key factor responsible for growth in the south of Germany and France. The latter represent the traditional lower-density peripheries of these two countries, whereas the faster growth in the UK's south covers both the south-eastern core of the national economy centred on London and the relatively low-density regions of the south west, East Anglia and rural Wales. In comparison, the cases of Portugal and Greece are quite straightforward with clear signs of further metropolitan concentration focusing on Lisbon, Oporto, Athens and their surrounding areas.

One very simple way of making sense of this information is to establish whether these patterns of population change are leading to greater regional concentration of population or whether they involve a process of dispersion towards a more uniform distribution. In other words, is there a positive or negative relationship between regional rates of growth and population density? Comparison of figures 3.1 and 3.3 confirms the concentration tendency in Portugal and Greece. Table 3.5 gives correlation coefficients for the association between the annual average rate of population increase and the log of population density in 1988. Only Germany appears to have experienced significant inter-regional population deconcentration. The results for Belgium, Denmark, Italy, Spain and the UK also suggest deconcentration, but the relationships are not significant at the 5 per cent level, at least not for the regional framework adopted here, nor are the low concentration relationships for Eire and the Netherlands. Across the entire Community the picture is certainly unclear, but the results in table 3.5 do indicate a significant contrast between the two halves of the EC with the north characterized by regional deconcentration of population and population concentration prevailing in the south.

Faced with the rather varied nature of this evidence, researchers have attempted to formulate new hypotheses. One approach relates to the concentration–deconcentration dimension, attempting to set recent developments into a longer-term framework by stressing the importance of temporary factors or period effects. Here the major concern is whether it is the 1970s that should be considered an anomalous period in the continuing trend towards greater concentration or whether global economic recession has temporarily stalled a new general process of counter-urbanization. A second group of hypotheses gives special attention to the processes of economic restructuring, pointing to the long-term decline of employment in primary and manufacturing sectors in the developed world and the rapid growth of jobs in the

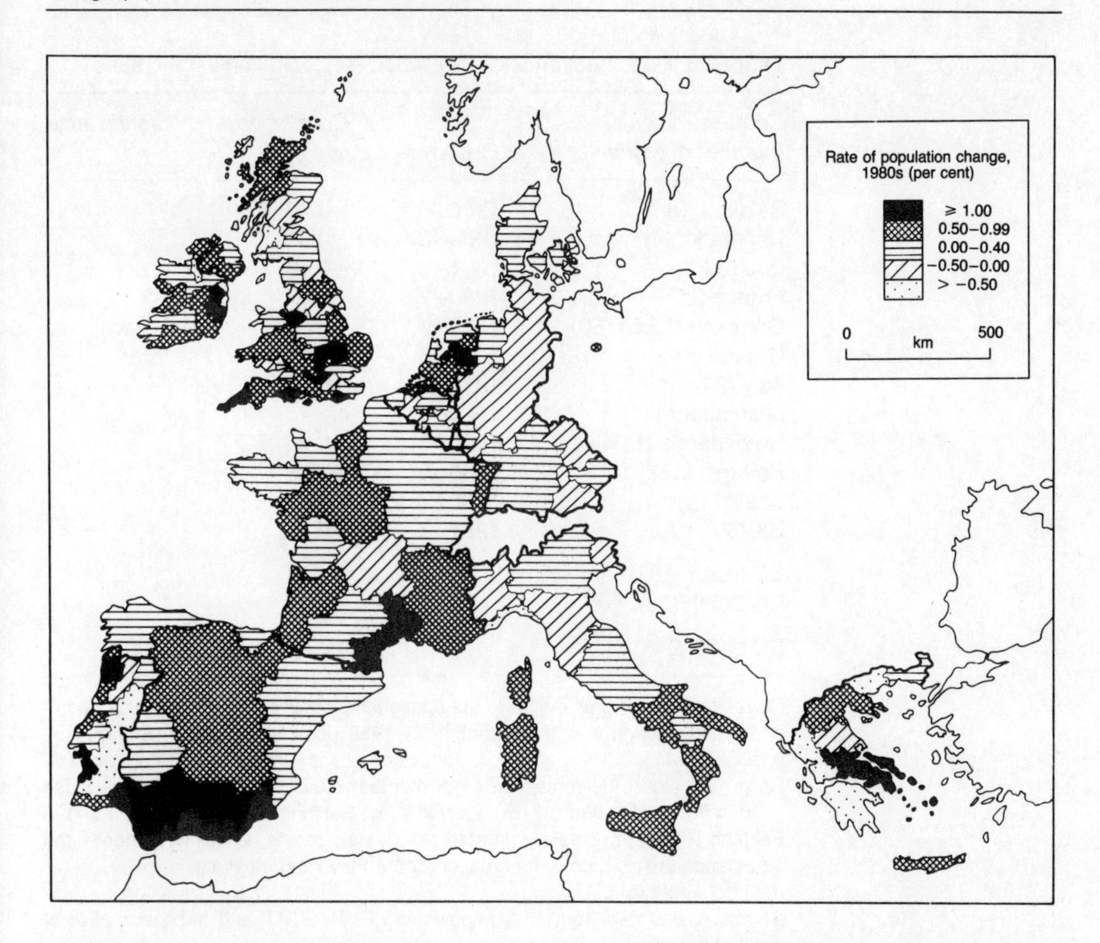


**Figure 3.3** Regional variations in annual average rate of population change in the 1980s.

'information economy'. The latter is said to favour the large metropolitan centres specializing in 'command and control' activities and the smaller-city 'prestige environments' attractive to key staff in research, development and other high-level jobs. A third approach portrays the regional map of Europe as a 'mosaic of dynamics and crisis', emphasizing the weakening of traditional core–periphery distinctions and pointing to a marked diversity of experience between superficially similar places. It is suggested that subtle differences in the 'chemistry' of localities determine whether they attract or repel the roving capital investment needed for economic growth.

In these circumstances it is not surprising that the more detailed forecasts of Europe's changing population geography are either based on forward projections of recent trends or represent 'what if?' exercises which explore the implications of adopting certain sets of assumptions about the most significant dimensions of

**Table 3.5** Regional trends in population concentration in the 1980s

| *Country (number of regions)* | *Period* | *Correlation coefficient* | *Significance level* |
|---|---|---|---|
| Belgium (9) | 1980–8 | −0.40 | |
| Denmark (14) | 1980–8 | −0.35 | |
| Eire (26) | 1981–6 | +0.16 | |
| France (22) | 1980–7 | +0.07 | |
| Germany (FRG) (30) | 1980–7 | −0.64 | *** |
| Greece (9) | 1980–8 | +0.78 | ** |
| Italy (20) | 1980–8 | −0.07 | |
| Luxembourg (1) | 1980–8 | n.a. | |
| Netherlands (11) | 1980–8 | +0.06 | |
| Portugal (18) | 1980–8 | +0.68 | ** |
| Spain (16) | 1980–7 | −0.17 | |
| UK (69) | 1981–8 | −0.15 | |
| EC North (182) | – | −0.19 | ** |
| EC South (63) | – | +0.37 | ** |
| EC of 12 (245) | – | −0.06 | |

Correlation coefficients measure the relationship between annual average rate of population change and the logarithm of 1988 population density. Significance levels: ***0.1%, **1%, *5%. EC South comprises Greece, Italy, Portugal and Spain; EC North the remaining eight member states. Regional units are those used in figures 3.1 and 3.3, except that West Berlin is omitted from FRG and in Belgium Brabant/Brussels is treated as a single region; in the Netherlands the boundaries are as before the creation of the Flevoland province.
n.a., not applicable.
*Source*: Calculated from data supplied by EUROSTAT and national statistical agencies

likely future trends. The work of Rees et al. (1992) is particularly relevant in this respect. They have projected forward the population of NUTS 1 regions to the year 2020 on the basis of two special scenarios. One, called 'growth regions', is based on assigning net migration performance to regions in direct proportion to their current gross domestic product (GDP) per capita rankings, while the other assumes that net migration is related to population density. The latter 'counter-urbanization–urbanization' scenario incorporates a distinction between northern countries where high density is linked to net out-migration and southern Europe (defined to include Eire and Northern Ireland) where the highest density regions are assumed to attract the strongest net in-migration. The population levels predicted by these two 'what if?' scenarios are then compared with a projection based on the combination of actual recent migration trends and one based on the assumption of no intra-EC migration at all.

The general patterns arising from these types of projection exercises will, of course, hold few surprises for those familiar with the geography of recent migration trends, patterns of population density and GDP per capita. The most interesting aspect of this work is the mechanism it provides to assess the extent to which assumptions are altered in the wake of further analyses of the 1980s trends and, no doubt, as a result of the reinterpretation of the 1970s experience in the light of more recent discoveries.

Since the increasing importance of intra-Community migration cannot be doubted in the 1990s and beyond, how may regional policies shape future events? The completion of the single European market at the end of 1992 could set in train a process of spatially concentrated economic growth that will draw increasing numbers of labour migrants into the more centrally located regions of the 'golden triangle' or 'blue banana', or perhaps – if it does – strong policy measures will be used to reinvigorate the peripheral regions. With or without the formal admission into the Community of certain republics from the Commonwealth of Independent States and other states in eastern Europe, will the westward flow of people from these countries take on again the dramatic dimensions that is assumed in 1988–90? In what ways and how effectively will the EC and its member states react to the continuing build-up of population pressure along its southern borders, including the mass movement of refugees and asylum-seekers from places like the former Yugoslavia, northern Africa and the Middle East?

These are all important issues which will affect the future distribution of population within Europe. At the macro-level, differentials in economic success are likely to condition rates of growth or decline, but these will also be affected by external political forces, national interests and social values. The future remains uncertain; the past tantalizingly obscured.

# 4

# Spatial Inequalities in Mortality

Daniel Noin

Research on the spatial inequalities in mortality within the European Community (EC) is still limited. There are a few studies on certain countries, but almost none on the Community as a whole. Geographical aspects of mortality variations have hardly been considered so far. How wide are the inequalities? How have they evolved? What are the main trends? What are the prospects for the year 2000? This chapter is directed towards these questions.

## An Assessment of the Situation

There seems to be general agreement that the level of life expectancy at birth in the Community is both high and satisfactory. This is certainly true when one considers only the industrial countries, but the matter is less obvious when the world's most favoured nations are used as a yardstick (table 4.1). Japan is undeniably ahead in both infant mortality and life expectancy at birth. The EC figures are good, but not the best. No Community country can be seen in the top rank for life expectancy at birth and several countries are far down the list. Although the poor ranking of Portugal and the former German Democratic Republic (31 and 28 in the world ranking) is understandable, that of the UK and the Federal Republic of Germany (22 and 20, respectively) is more difficult to explain considering the substantial investments in health services, social provision and education that have been made.

When one begins to examine sub-national variations in mor-

**Table 4.1** Comparative mortality measures

| | *Infant mortality rate (per 1,000)* | | *Life expectancy at birth (years)* | |
|---|---|---|---|---|
| | *1980–5* | *1988–9* | *1980–5* | *1988–9* |
| Japan | 7 | 5 | 76.5 | 79.0 |
| Scandinavia | 8 | 7 | 75.7 | 76.7 |
| Australia/New Zealand | 10 | 8 | 74.9 | 76.0 |
| European Community | 11 | 8 | 74.3 | 75.8 |
| USA/Canada | 11 | 9 | 74.5 | 75.5 |

tality, distinctive and systematic inequalities emerge. The differences are more accentuated in the figures for infant mortality, as figure 4.1 shows (Decroly and Vanlaer, 1991). The advance of the Netherlands, Denmark and France, with universally low infant mortality rates, is quite clear, whilst Portugal is lagging behind. Other countries show sharp internal contrasts, especially Germany, Spain, Italy and Great Britain. The extreme differences are quite considerable with variations ranging from one to five or more. The lowest figures are to be found in certain Danish districts around Copenhagen and the highest in parts of northern Portugal (see also figure 4.3).

For life expectancy at birth, the differences are not very marked when national figures are considered. The maximum difference between countries in 1980 was 4.4 years (figure 4.2). The Netherlands and Portugal again occupied the first and last places on the list. By the end of the 1980s Portugal's place had been taken by the German Democratic Republic.

Of course, the inequalities are more pronounced if finer spatial divisions are considered (figures 4.3, 4.4 and 4.5). (Note that the estimates of life expectancy at birth for men and women in figures 4.4 and 4.5 have been derived from standardized mortality indices (see Poulain, 1990).) For women in 1980, life expectancy at birth was already over 80 years in some geographical units, especially in northern Spain, while it was less than 75 in many districts of East Germany and Portugal. For men, for the same year the equivalent figures are over 73 and under 68 years. The effects of national policies are quite evident. Some countries are markedly distinct from their neighbours. East Germany and Portugal attract attention because of their generally high levels of mortality whereas Denmark, the Netherlands, France and Spain display rather better levels. Internal inequalities are clearly marked in the UK where the

**Figure 4.1** Trends in infant mortality rates.

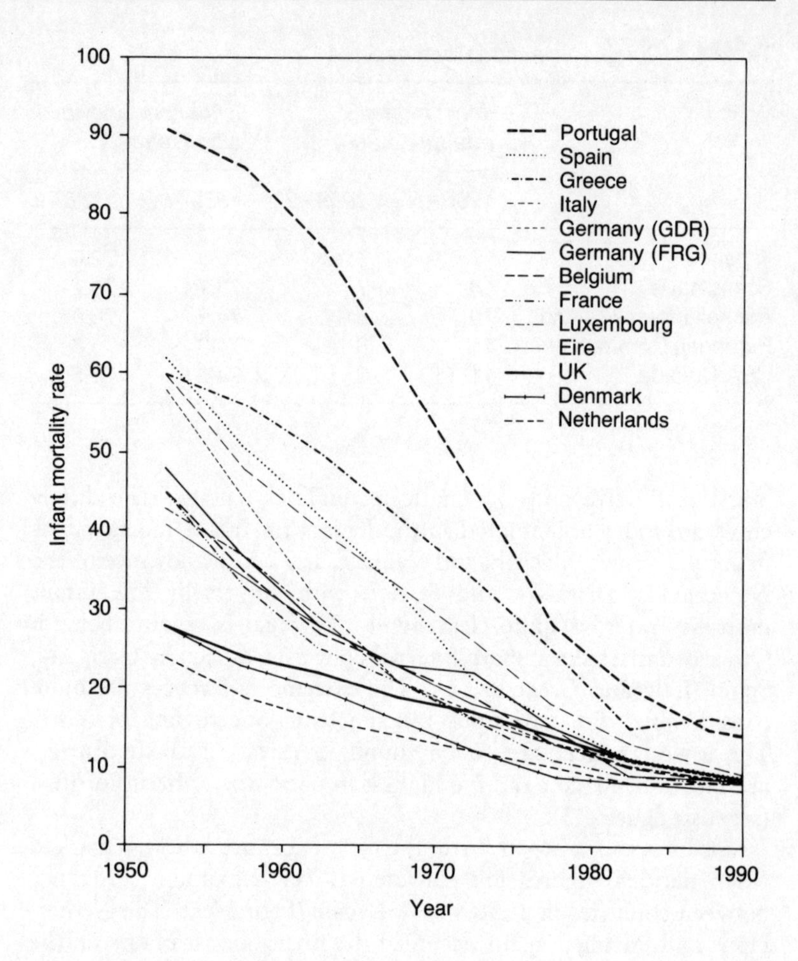


south east is contrasted against the rest of the country. In Belgium, the classic opposition between Flanders and Wallonia still persists in terms of mortality patterns. In Italy, there are regional differences between the north and the south, but in this case it is the south that has higher life expectations.

The factors influencing mortality are numerous and their interaction is complex. To date there has been no comprehensive study of the geographical variations in mortality levels throughout Europe, but national studies indicate that the following are points at issue. The relationship between life expectancy at birth and the average level of wealth is not strong. Luxembourg has the highest gross national product (GNP) per capita in the Community but ranks only tenth in terms of life expectancy at birth, while Spain is in second position in terms of life expectancy but has a GNP per capita almost three times lower. The association between mor-

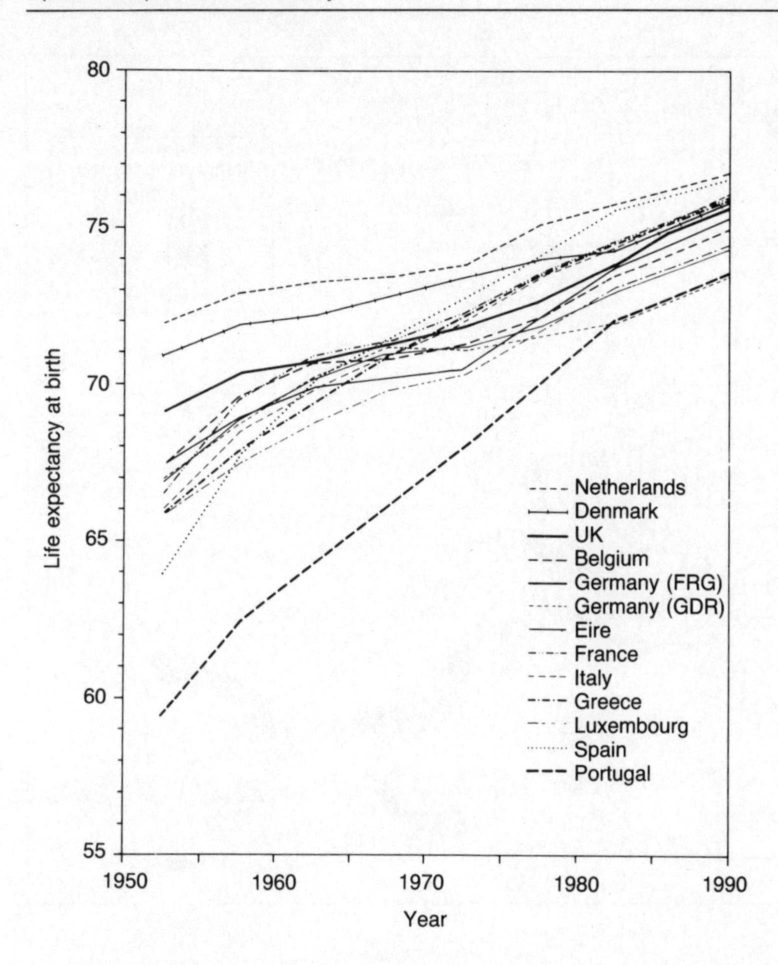


**Figure 4.2** Trends in life expectancy at birth.

tality level and the quality of health care and social provision is also not as clear cut as one might have expected (Kunst et al., 1988). Several Mediterranean countries have relatively poorly developed services in this area, but they also have relatively low levels of mortality. The relationship with the average level of education does not seem particularly strong; in this respect also, the findings from southern European countries are puzzling.

It would now appear that the essential factors are linked to behaviour, in particular to smoking and the consumption of alcohol and animal fats which are generally at higher levels in the northern countries of the Community than in the southern countries. For example, death from cancer of the digestive system is undeniably linked to alcohol consumption, being six times greater in north-western France than southern Italy. Mortality from lung cancer is high in the UK, the Benelux countries and

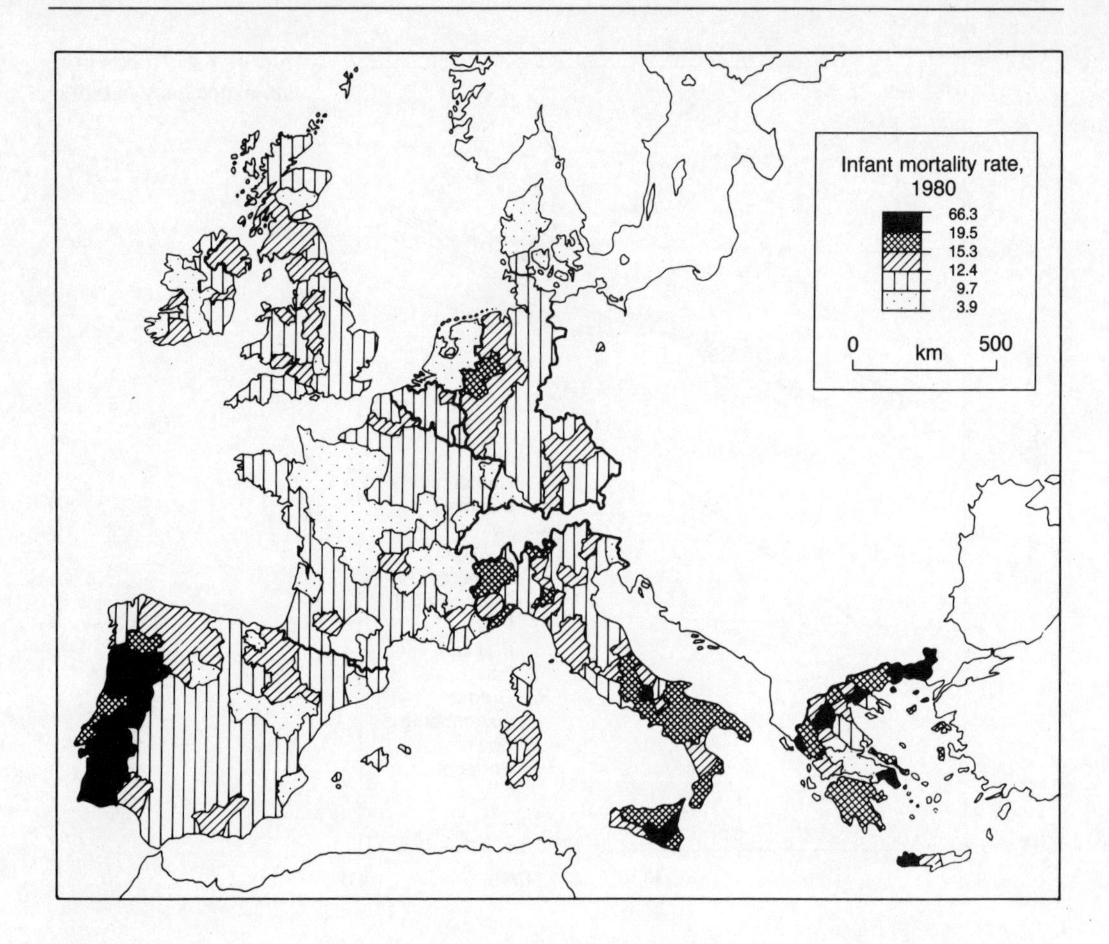


**Figure 4.3** Regional variations in infant mortality rates, 1980.

Germany where cigarette smoking was widespread and started early among young people at least a generation ago. Death from cardiovascular diseases is much more significant in countries with a high consumption of animal fats – for example in the British Isles, Belgium, Germany and northern France – than in those countries where food is generally cooked in olive oil, as in the Mediterranean regions. It has also been observed that mortality from cancer and cardiovascular diseases is lower in southern Europe where fruit and vegetable consumption is traditionally higher.

Further research on the relative influence of these factors would doubtless be assisted by the construction of more detailed mortality maps and the development of a distinct spatial perspective (see Holland, 1988).

There remain strong and persistent differences between the life expectancies of men and women (table 4.2 and figures 4.6 and 4.7). Male mortality is higher at all ages in the EC, but especially

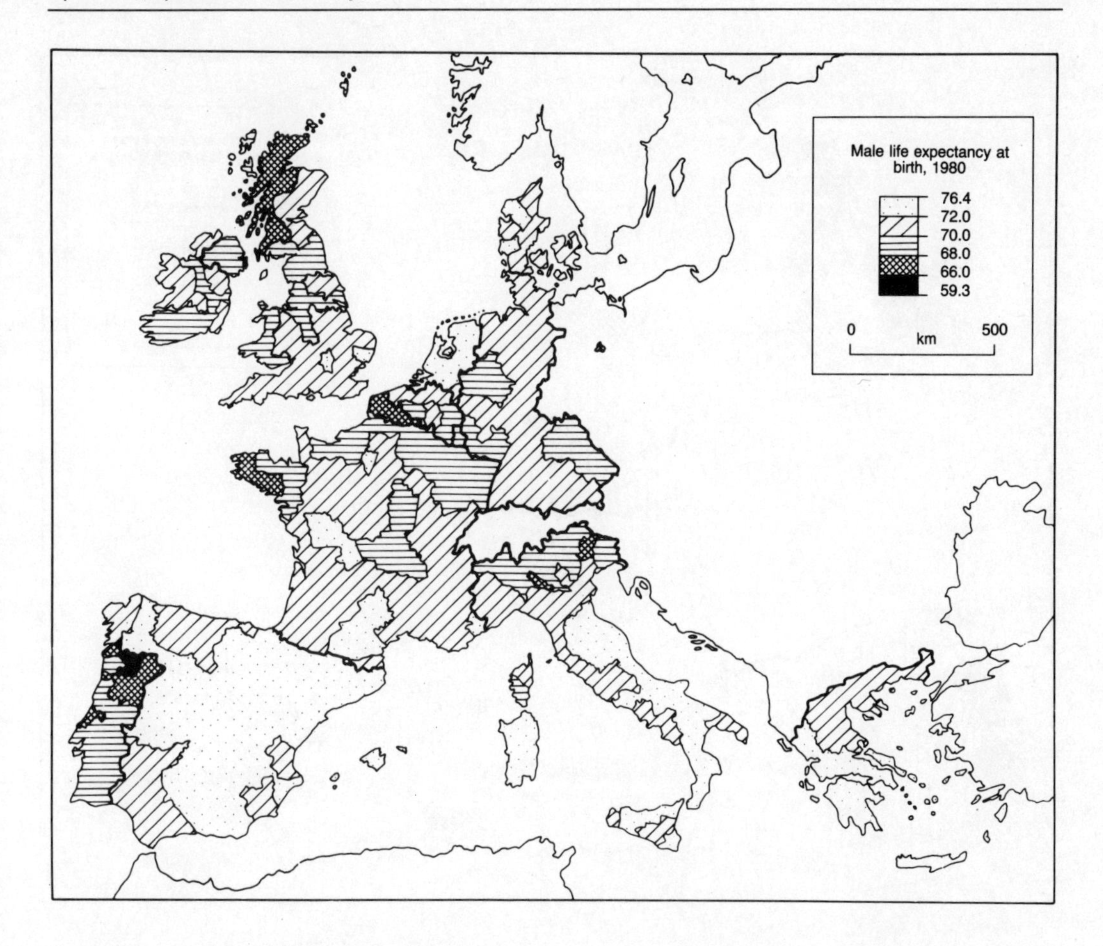


Figure 4.4 Regional variations in male life expectancy at birth, 1980.

among young men and older adults. For the young, accidents and violent deaths are largely responsible for the difference, while for adults cancer and cardiovascular diseases are more important. The last mentioned are particularly associated with gender-specific behavioural differences with respect to smoking, alcohol consumption and diet.

Several authors have considered the effects of different smoking habits (Hart, 1986). Smoking no doubt accounts for several years of difference in life expectancy at birth between men and women in the northern part of the Community, where it used to be more widespread 20 or 30 years ago and where people started smoking at an earlier age. Alcoholism is undoubtedly responsible for part of the difference in certain regions of Europe such as north-western France, northern Italy and the northern margins of Spain. Variations from one country to another in these mortality differentials are in fact quite significant (table 4.3).

Closer inspection reveals that these gender-specific differences

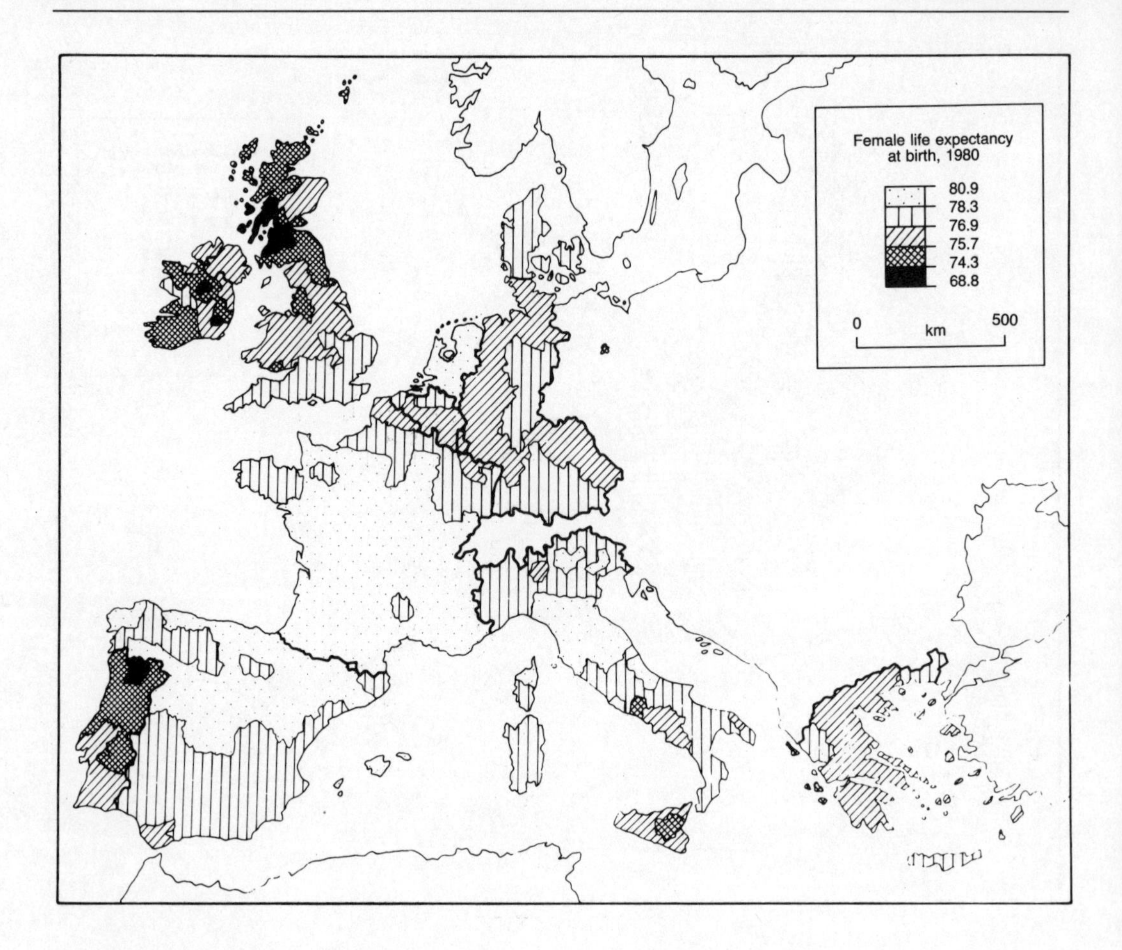


**Figure 4.5** Regional variations in female life expectancy at birth, 1980.

are even more pronounced (figure 4.7). The greatest differences in life expectancy at birth are found in the French and Italian districts where the gap may be as much as ten years in some places (Finisttèrre, Morbihan and Belluno for instance). But the difference is less than four or five years in some other parts of Italy and in Greece (Sicily, the Ionian Islands, the Peloponnisos and Crete).

Mortality differences between social groups are far more difficult to observe and compare. There are severe definitional problems in distinguishing socio-professional groups and developing standardized classifications of occupations even for one country, but these problems are compounded if one wishes to compare the situations and trends in different European countries.

It seems from the literature available that social inequalities are quite significant, especially for men, and that they do not appear to be disappearing, but are both highly persistent and resistant to developments in modern medicine and health care. Those socio-

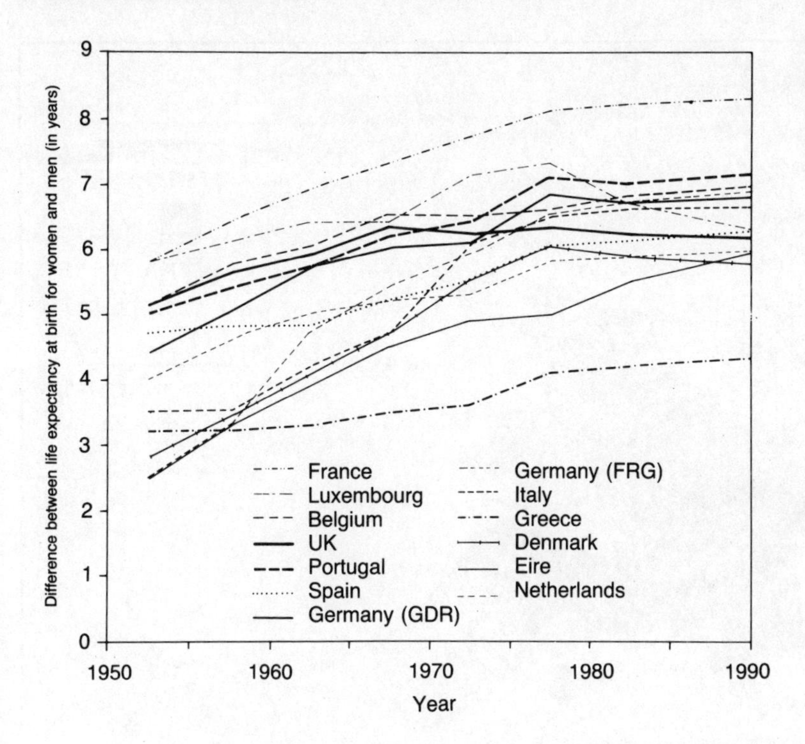


**Figure 4.6** Trends in gender differentials in life expectancy at birth.

**Table 4.2** Differences in life expectancy between men and women (years in 1980–5)

| | |
|---|---|
| Japan | 5.5 |
| Scandinavia | 6.1 |
| Australia/New Zealand | 6.6 |
| European Community | 6.7 |
| USA/Canada | 7.4 |

professional groups that are most favoured by longevity are company executives and members of the liberal professions, whereas unskilled manual workers are found to have the highest levels of adult mortality. It is also found to be the case that, whereas the differentials between all other groups are narrowing, that between all others and the unskilled group is widening. Social inequalities in mortality patterns seem to be more closely linked to the level of education and training and to life-styles, rather than simply to the level of income (Desplanques, 1984; Wilkinson, 1986; Valkonen, 1987; Fox, 1989).

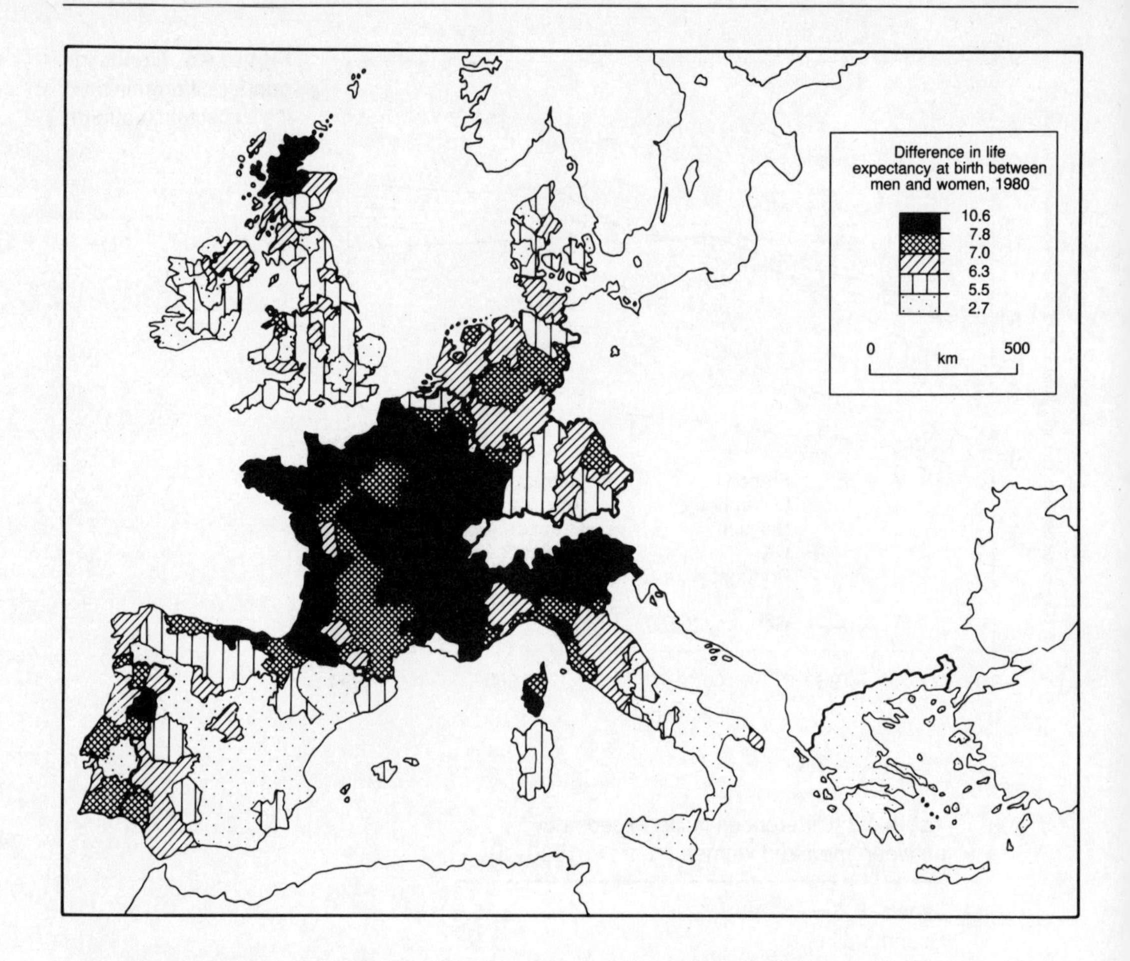


**Figure 4.7** Regional variations in gender differentials in life expectancy at birth, 1980.

**Table 4.3** Differences in life expectancy between men and women in the European Communily countries (years)

| | | | |
|---|---|---|---|
| Eire | 5.5 | Luxembourg | 6.7 |
| UK | 6.2 | France | 8.2 |
| Denmark | 5.9 | Portugal | 7.0 |
| Germany (FRG) | 6.7 | Spain | 6.1 |
| Germany (GDR) | 5.9 | Italy | 6.6 |
| Netherlands | 6.7 | Greece | 4.2 |
| Belgium | 6.8 | | |

**Recent Evolution and Trends**

There has been a considerable decline in infant mortality over the last four decades averaging out at a reduction of about 5 per cent

**Table 4.4** Evolution of life expectancy in the European Community (years)

| | *1950–5* | *1970–5* | *1980–5* | *1988–9* |
|---|---|---|---|---|
| Men | 64.7 | 68.8 | 71.2 | 72.6 |
| Women | 69.2 | 74.9 | 77.8 | 79.3 |
| Difference | 4.5 | 6.1 | 6.6 | 6.7 |

per year. Its level is now four times less than it was in 1950 (figure 4.3).

Life expectancy at birth has also increased substantially since 1950, by about nine years on average or approximately three months a year. However, this progress has not been regular; it was rapid in the 1950s and early 1960s, slower in the late 1960s and early 1970s and it has accelerated again since the late 1970s. Overall, progress has been steady (figure 4.2). The increasing success of the campaign against smoking, the decrease in the consumption of alcohol and animal fat, together with a better level of education and a more frequent practice of physical exercise are all factors that are likely to prove favourable in the years to come. (See chapter 8 for a detailed consideration of mortality differentials at later ages, especially tables 8.6–8.8.)

Although trends are favourable as a whole, there are quite clear differences in the experiences of the two sexes. Since the early 1950s, life expectancy at birth in the Community has increased by 10.1 years for women, but only 7.9 years for men (table 4.4). The difference in longevity between the sexes has increased significantly, but the rate of its change has appeared to slow down recently. If the figures for 1988–9 are confirmed, there will be good reason to expect that the gap will begin to narrow in the future as the consumption habits of men and women become more alike, especially in terms of smoking.

Immediately after the Second World War mortality differences between the European countries appeared very marked. Some national populations experienced mortality levels well below the EC average, notably the Netherlands and Denmark, while others were well in excess, for example Spain and Portugal. These international differences have markedly diminished as those countries that formerly lagged behind have made rapid and remarkable progress. Between the early 1950s and the early 1980s life expectancy at birth increased by 4.4 years in Denmark and 14 years in Portugal.

Hence the process of convergence is quite obvious: international and inter-regional differences have been much reduced and will probably continue to be so in the near future. Estimates of the maximum possible average life expectancy at birth vary slightly, but 92 years would appear to be a good approximation. The rate of change of European mortality will probably slow even further as this upper limit is approached, spatial variations will narrow even further, as may those between men and women, but the social differences are likely to persist as the clearest element of differentiation within the overall mortality pattern.

# 5

# Fertility: the Lowest Level in the World

Deborah Sporton

'Below replacement' and 'birth shortages' are both terms that have been associated with the dynamics of European fertility during the post-Second World War period (Davis et al., 1987). The general trend of fertility decline reflects a continuation of the momentum generated during the fertility transition which began in some countries more than a century ago. The general downward trend, however, has been interrupted by cyclical upturns during the last half-century. The European fertility transition was characterized by considerable temporal and geographical diversity. Today, however, all countries in western Europe have fertility either converging towards, or already at, levels which are no longer sufficient to allow natural population growth.

There is no prevailing consensus of opinion on the causes of these trends. Some believe that the social norms and values governing childbearing have changed. Others, notably those who are influenced by theories current in micro-economics, argue that it is the very economic structure of society that has changed, thus constraining the fulfilment of otherwise dominant behavioural norms.

This chapter examines the spatio-temporal variations in European fertility and their immediate or proximate determinants before focusing upon the contributory economic and social influences engendering the lowest levels of fertility in the world.

## Post-war European Fertility Trends: Below-replacement Fertility

The term 'below-replacement fertility' describes the inability of a population to grow naturally and is contingent upon a shortage of live births compared with the prevailing level of mortality. In the economically developed and medically advanced societies it is commonly associated with having a total fertility rate (TFR) of less than 2.1. The TFR is the average number of children a woman would be expected to have during her reproductive life if she conformed to the prevailing age-specific fertility rates. In practice, migration will often supersede these natural influences in determining absolute changes in population size.

A corollary of the temporal leads and lags associated with the fertility transition in Europe was the presence of both inter- and intra-national fertility differences (Coale and Watkins, 1986). The fertility transition, complete in most north-west European countries by the 1930s, was only concluded within the last few decades in many southern European countries. Today, all countries, with the sole exception of Eire, are experiencing below-replacement fertility and some regions, for example Emilia-Romagna and Liguria in northern Italy with a TFR less than 1.5, have very low fertility indeed (Rallu, 1983). Even Eire, which has traditionally maintained higher levels of marital fertility than its neighbours, has experienced a halving of its TFR from more than 4 to just over 2 during the 1980s (table 5.1 and figure 5.1).

Below-replacement fertility has emerged as a characteristic feature of post-industrial society and is therefore not a phenomenon restricted to Europe. Other economically developed regions of the world are today experiencing low levels of fertility, but they are still generally slightly higher than in Europe. For southern European countries, current low fertility levels set a historical precedent; among north European countries fertility was already at sub-replacement levels during the 1930s. The inter-war years were therefore characterized by distinct geographical variations in

**Table 5.1** Total fertility rates, 1990

| | |
|---|---|
| Belgium | 1.6 |
| Denmark | 1.7 |
| Germany | 1.5 |
| Greece | 1.5 |
| Spain | 1.3 |
| France | 1.8 |
| Eire | 2.2 |
| Italy | 1.3 |
| Netherlands | 1.6 |
| Portugal | 1.4 |
| UK | 1.8 |

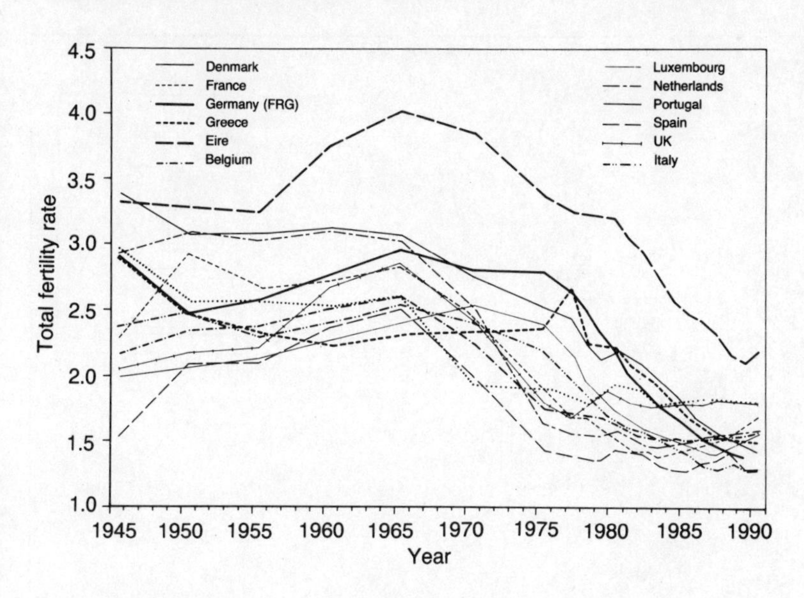


**Figure 5.1** Trends in total fertility rates.

fertility which manifest themselves in terms of a north–south or even a core–periphery division.

Immediately after the Second World War, there was a resurgence of births in northern Europe which has often been attributed to the compensatory effects of married couples reuniting or the formation of new permanent sexual unions delayed by the war. The 'baby boom' was to continue until the mid-1960s and would therefore appear to owe more to fundamental social and economic changes than merely to compensatory effects alone. These particular influences will be examined in due course. The post-war baby boom was noted for its contemporaneity in virtually all the countries of northern Europe. Geographical variations between individual countries which existed in the 1930s were also reduced, suggesting a polarization of fertility decision making among sexually active couples (figures 5.2 and 5.3). From the mid-1960s, fertility levels fell systematically throughout northern Europe and, with the exception of certain minor recoveries, have remained at below replacement level ever since. In southern Europe fertility decline began rather later.

The decline of fertility has been associated with a second demographic transition (van de Kaa, 1987). Classical transition theory incorporated a temporal sequence of events and yet failed to envisage the dynamics of population change following the adjustment from high to low mortality and fertility levels beyond the transition itself (Campbell, 1974). As biological and physiological maxima are approached in the medical quest to increase

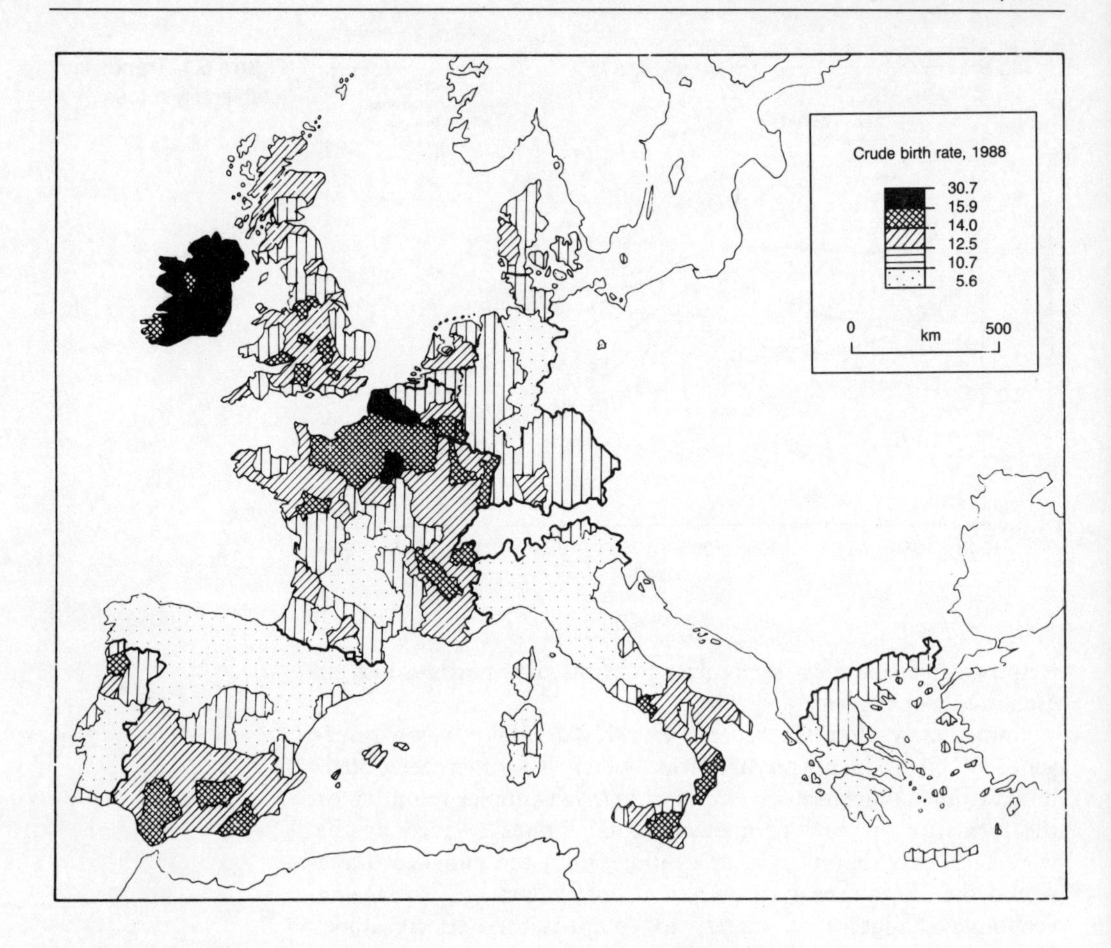


**Figure 5.2** Regional variations in crude birth rates, 1988.

longevity, cyclical fluctuations in fertility have become the driving force behind alterations in rates of natural population change in post-industrial, post-transitional societies.

### The Proximate Determinants of Low and Fluctuating Fertility

The decline in European fertility has been accompanied by significant changes in reproductive behaviour through which the influence of the broader social and economic environment is mediated. These proximate causes of European fertility decline include a change in the timing and age distribution of childbirths and more effective regulation of completed family sizes. Post-war fertility trends reflect, in part, changes in the timing of childbirths. In particular, fertility decline in northern Europe since the mid-1960s may be associated with the postponement of childbearing.

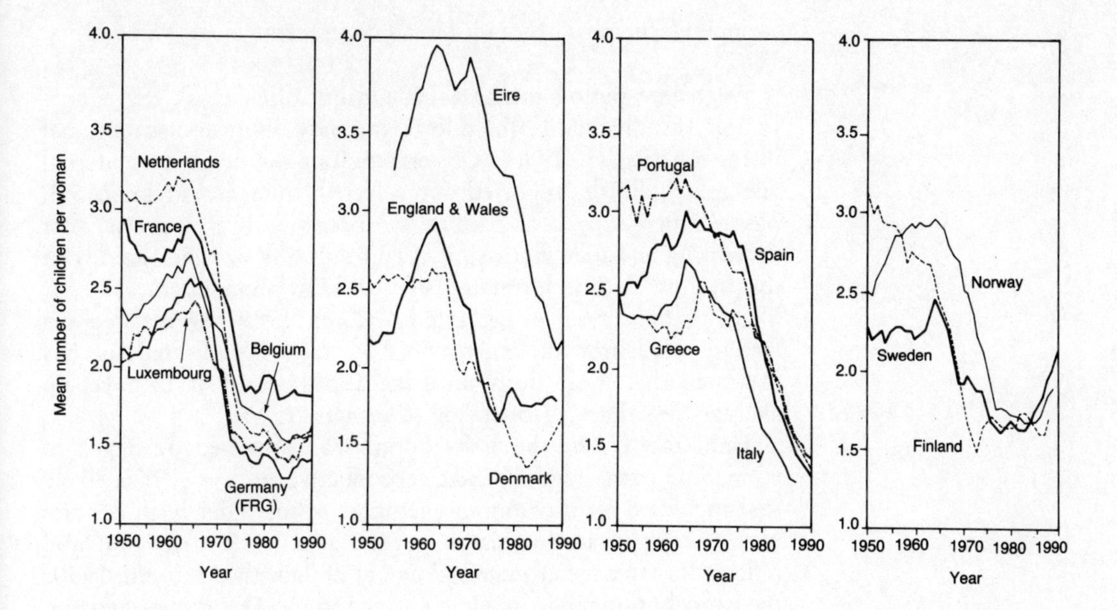


**Figure 5.3** Trends in the numbers of children per woman by years.

**Table 5.2** Mean age at first birth

| *Country* | *1970* | *1975* | *1980* | *1985* | *1988* |
|---|---|---|---|---|---|
| Belgium | 24.3 | 24.2 | 24.5 | 25.5 | n.a. |
| Denmark | 23.7 | 24.0 | 24.6 | 25.5 | 26.1 |
| FRG | 24.3 | 24.8 | 25.2 | 26.2 | 26.7 |
| Greece | 24.0 | 23.6 | 23.3 | 23.7 | 24.2 |
| Spain | n.a. | n.a. | 24.6 | 25.4 | n.a. |
| France | 23.8 | 24.2 | 24.9 | 25.2 | n.a. |
| Eire | 25.3 | 24.8 | 24.9 | 25.6 | 26.0 |
| Italy | 25.0 | 24.7 | 24.4 | 25.1 | n.a. |
| Netherlands | 24.3 | 25.0 | 25.6 | 26.5 | 27.2 |
| Portugal | 24.4 | 24.0 | 23.6 | 23.8 | 24.3 |
| UK | 23.9 | 24.6 | 25.1 | 25.9 | 26.6 |

n.a., not available.
*Sources*: Council of Europe, 1990a

Table 5.2 shows that by 1988 the average age of a woman at the birth of her first child was 26 years in northern Europe. Italy, Spain, Portugal and Greece had an earlier maternal age at first birth, but even here there are signs of a move towards postponement. Low southern European fertility levels are therefore engendered not so much by delayed 'starting' behaviour but by earlier 'stopping', the result of widespread contraceptive use (van de Kaa, 1987).

Aggregate period measures of fertility often mask the contribution of individual birth cohorts and may result in distributional distortion (Ryder, 1980). Cohort analysis of the mean maternal age at childbirth, reported for selected countries in table 5.3, has revealed a cyclical pattern in successive birth cohorts, first decreasing and then increasing. As a result, the size and rapidity of the fertility decline estimated by the TFR, a period measure, is likely to have been exaggerated (Prioux, 1989). Overall cohort fertility measures do mirror period trends and analysis has indicated that those born since the 1950s are likely to have, on average, less than 2.1 offspring (Chesnais, 1990).

Until the 1950s, the baby boom was sustained by births to women at each stage of their reproductive life-cycle. It is likely that increased fertility among members of the older birth cohorts was to compensate for births forgone as a result of the war and reflected a later mean maternal age at childbirth. From the 1950s the contribution made by older women to the TFR diminished and the baby boom was supported thereafter by births to younger women. From the mid-1960s, although the fertility of women at all ages fell, the most significant reductions were initially to women within the older age groups (Campbell, 1974). Since 1975 the fertility of older women has stabilized and in some countries, for example the UK, France and Germany, there has been an upturn while births to younger women have been reduced so that overall fertility decline has been forestalled (Muñoz-Perez, 1986; B. Werner, 1988). Post-war fertility fluctuations may thus be associated with changes in the age distribution of women at childbirth.

**Table 5.3** Mean age at childbirth according to birth cohort

| | *Birth cohort* | | | | |
|---|---|---|---|---|---|
| | *1935* | *1940* | *1945* | *1950*[a] | *1955*[a] |
| Denmark | 26.2 | 25.7 | 25.6 | 26.1 | 26.8 |
| France | 27.1 | 26.5 | 26.0 | 26.4 | 26.8 |
| Belgium | 27.2 | 26.4 | 25.9 | 26.1 | 26.2[b] |
| Netherlands | 28.1 | 27.1 | 26.5 | 27.0 | 27.6[b] |

[a] Estimates.
[b] 1954 birth cohort.
*Source*: Prioux, 1989, p. 162

This trend towards a later average age at first birth may be associated with postponement of marriage as the majority of births still take place within marriage. Changes in the proportions of married women particularly within the younger reproductive age groups may also be allied to post-war fertility fluctuations in Europe. van de Kaa (1987, p. 10) has linked declining European fertility rates to the transition from 'the golden age of marriage to the dawn of cohabitation'. But this association does not help us to account for the 'baby bust' which took place during the late 1960s at a time of near universal marriage (see chapter 9).

A further significant proximate cause of low European fertility has been the tendency towards greater limitation of births through spacing and stopping behaviour. The fertility decline may be associated with the emergence of a modal family comprising at most two children (Berent, 1983).

Table 5.4 clearly indicates a decline in the representation of fourth and higher order births. In most European countries first- or second-order births now account for approximately 80 per cent of all live births. There is some evidence to suggest a recent decline in the proportions of women having higher order births (Prioux, 1989). Improved appliance methods of contraception, sterilization, wider access to legal abortion and the introduction of oral contraceptives have all contributed to smaller family sizes by reducing the number of unwanted pregnancies. Increased voluntary childlessness may be supplemented by the effects of lower fecundity as women delay childbearing until later in their reproductive lives, as well as marital dissolution (Kiernan, 1989, 1992). Voluntary childlessness is also likely to be of increasing importance in the future (Lesthaeghe, 1983; Humphreys, 1988).

## The Broader Determinants of Contemporary Fertility Fluctuations

Despite the voluminous literature which considers the economic, social and even political determinants of low and fluctuating fertility, there is no single Europe-wide interpretation and if one were to be proposed it would surely prove far too simple. However, certain strands of the argument do already exist. Considerable effort has been expended in ascribing fertility levels to economic causes (for example, Becker, 1960; Silver, 1965; Leibenstein, 1974; Easterlin, 1976, 1980). In summary, children are viewed as consumer durables and fertility decision making is thought to involve rational consideration of the cost of child rearing in relation to overall household income and expenditure. In times of economic recession, therefore, when resources are

**Table 5.4** Percentage live births according to birth order

| | *1st birth* | *2nd birth* | *3rd birth* | *4th+ birth* |
|---|---|---|---|---|
| *Belgium* | | | | |
| 1975 | 49.1 | 30.8 | 11.1 | 9.0 |
| 1980 | 47.9 | 39.3 | 12.1 | 7.1 |
| 1985 | 46.4 | 33.7 | 12.8 | 7.1 |
| 1986 | 46.2 | 33.2 | 13.3 | 7.2 |
| *Denmark* | | | | |
| 1975 | 43.4 | 33.3 | 14.2 | 9.1 |
| 1980 | 45.7 | 35.4 | 14.2 | 4.7 |
| 1985 | 38.1 | 37.2 | 18.6 | 6.1 |
| 1988 | 46.9 | 36.8 | 12.4 | 3.9 |
| *FRG* | | | | |
| 1975 | 46.7 | 33.0 | 11.8 | 8.5 |
| 1980 | 48.7 | 34.3 | 11.1 | 5.9 |
| 1985 | 48.4 | 35.7 | 11.3 | 4.6 |
| 1988 | 47.8 | 35.3 | 12.0 | 4.9 |
| *Greece* | | | | |
| 1975 | 43.4 | 36.8 | 13.0 | 6.8 |
| 1980 | 44.9 | 37.4 | 12.7 | 5.0 |
| 1985 | 41.2 | 39.0 | 12.0 | 4.8 |
| 1988 | 45.4 | 38.5 | 11.4 | 4.7 |
| *Spain* | | | | |
| 1982 | 43.4 | 31.7 | 14.3 | 10.5 |
| 1986 | 44.5 | 35.1 | 12.6 | 7.7 |
| *France* | | | | |
| 1975 | 48.2 | 31.0 | 11.3 | 9.5 |
| 1980 | 44.2 | 34.6 | 14.6 | 6.6 |
| 1985 | 42.1 | 34.9 | 14.9 | 8.1 |
| 1988 | 40.7 | 34.3 | 16.3 | 8.7 |
| *Eire* | | | | |
| 1975 | 30.7 | 24.4 | 17.3 | 27.7 |
| 1980 | 29.3 | 24.3 | 19.3 | 27.1 |
| 1985 | 32.1 | 25.2 | 18.1 | 24.6 |
| 1989 | 32.8 | 27.3 | 18.4 | 21.5 |
| *Portugal* | | | | |
| 1975 | 41.3 | 27.1 | 11.7 | 19.9 |
| 1980 | 45.4 | 31.3 | 11.0 | 12.3 |
| 1985 | 47.1 | 31.5 | 11.3 | 10.1 |
| 1988 | 49.9 | 31.4 | 10.2 | 8.5 |
| *UK* | | | | |
| 1975 | 41.2 | 37.1 | 13.5 | 8.2 |
| 1980 | 41.3 | 35.9 | 15.1 | 7.7 |
| 1985 | 34.7 | 36.2 | 15.7 | 8.3 |
| 1988 | 40.4 | 35.8 | 15.5 | 8.3 |

*Source*: Council of Europe, 1990a

squeezed, fertility levels will be lower than during times of economic prosperity. Prevailing low European fertility levels may be associated with the current economic recession and cyclical fluctuations in fertility may be linked to economic cycles. The implication of such economic arguments is that fertility rates will not continue to decline indefinitely; instead, a future economic upturn will stimulate a demographic revival.

It has also been argued that fertility oscillations are self-perpetuating phenomena influenced by cohort size and relative economic status (Easterlin, 1980). When the immediate post-war birth cohort reached adulthood, the labour market experienced a supply surplus which depressed wage rates and adversely affected living standards in general. According to Easterlin, the decision to have a child will be influenced by material aspirations nurtured during adolescence. In the case of the 'baby boomers', this may be associated with a period of relative economic prosperity. If material aspirations are not met during adulthood there will be a distinct reluctance to raise large and thus costly families; the birth of high-parity children will be unlikely. Current low levels of European fertility may simply reflect the presence of a large birth cohort experiencing lower relative economic status than that encountered during adolescence. Conversely, the smaller birth cohorts of the 1970s and 1980s should enjoy increased relative economic status as they enter the labour market and, in theory at least, fertility should increase.

Easterlin's model takes no account of the significant increases in female labour force participation since the 1970s, which have undoubtedly influenced family incomes (Ermisch, 1990a). Although mother-intensive child rearing is now far less popular, the dual roles of mother and working woman create severe social and psychological tensions despite improvements in workplace child-care facilities and better maternity-leave provision in some western European countries (see chapter 6). The high opportunity costs associated with leaving the labour market to have children are likely to depress fertility levels. It is interesting to note that, in Sweden, a recent upturn in the TFR from 1.6 in 1983 to 2.1 in 1990 has been linked with social policy measures providing financial incentives for working women, or their partners, to take leave and to have not one but often two closely spaced births (Hoem, 1990). To date, Sweden has played the role of trendsetter with regard to European demographic developments; policies embodied in the European Social Charter may improve economic and social conditions for working mothers and extend the Swedish experience elsewhere.

However, Monnier (1990) has questioned the universality and

hence the significance of the relationship between fertility and female activity rates. The southern European countries of Italy, Spain and Greece, for example, displayed both the lowest European fertility levels and the lowest female activity rates in 1990. In Italy, Spain and Greece no more than 44 per cent of women participated in the labour force compared with 81 per cent of all Swedish women. Clearly, the significance of the relationship between female labour force participation and fertility varies geographically within Europe, suggesting the interplay of other factors.

Although the importance of economic influences upon European fertility levels should not be discounted, changes in reproductive behaviour during the post-war period may also be allied with a shift in social mores and values (Murphy, 1992). Within the West, the parental function is no longer deemed to be as socially gratifying as in the past. Attitudes have changed from the child-centred, altruistic values that characterized the first fertility transition in the nineteenth century to the more individualistic, post-materialist values of today. Low fertility levels prevailing at the end of the demographic transition were engendered by the parental desire to have fewer 'quality' children. Today, very low fertility may be associated with greater individualism as couples no longer plan their lives around the birth of a child and his or her future. The contemporary situation is aptly described by van de Kaa (1987), who contrasts the shift in importance from the 'king child' of the 1930s to the 'king pair' of today.

A further socio-cultural transformation that may be associated with the trend towards greater individualism has been that of secularization. Simons (1986) asserts that religious beliefs were important in shaping pro-natalist attitudes in western Europe during the fertile 1950s and 1960s. Low fertility levels over the last three decades are likely to have been influenced by the relegation of religion from the public to the private sphere. Previously, religious norms dominated all spheres of activity in both the public and private domains. Individuals are now able to choose more freely between competing value systems. In this regard, the traditional Roman Catholic strongholds of southern Europe have experienced the most dramatic reductions in fertility in recent years (Muñoz-Perez, 1987).

The increasing popularity of alternative living arrangements and the weakening of traditional family ties in response to more liberal societal attitudes have also indirectly promoted lower fertility. Consensual unions are now widely accepted; however, only in Norway and Sweden has cohabitation emerged as an alternative to, rather than a precursor of, marriage (Keilman, 1987; Hoffmann-Nowotny and Fux, 1991). In most European countries the majority of births are postponed and take place within the confines of wedlock (Moors and van Nimwegen, 1990). There is further evidence to suggest that the fertility of cohabiting couples is lower than that of their married counterparts (Leridon and Villeneuve-Gokalp, 1988). Monnier (1990, p. 73) has offered an explanation: 'It is rare for married couples to remain voluntarily without a child. The child had become one of the quasi-obligatory attributes of marriage. By contrast in the case of unmarried couples, the birth of a child, with exceptions, is not always one of the outcomes of such a union.'

The institution of marriage was once considered sacrosanct; today rising divorce rates suggest a shift in societal attitudes. As the majority of births still take place within marriage, increasing marital instability has also contributed to low European fertility levels by reducing the length of a woman's marital reproductive life (Kiernan, 1989), a situation exacerbated by the short duration of an increasing proportion of those marriages ending in divorce (Sardon, 1986) (see figure 9.2). Moreover, there is evidence to suggest that remarriage rates among divorcees are declining, further reducing the marital reproductive period (Festy, 1985; Ermisch, 1990b).

The proximate determinants of below-replacement fertility have been influenced by the complex interplay of social and economic forces. Despite the convergence, within Europe, of aggregate fertility levels, the relative importance of the broader societal determinants continues to vary spatially.

## Prospects for Fertility in Eastern Europe

Although this chapter has mainly been concerned with variations in fertility among the countries of the European Community, recent political developments in eastern Europe are likely to have profound implications for the population geography of the New Europe. It is therefore important to mention the fertility characteristics of these former communist bloc countries.

Fertility levels in the former eastern bloc countries, although marginally higher than those observed in the West, are barely sufficient to ensure the natural replacement of the populations (table 5.5). However, the proximate determinants of fertility are different; marriage rates are higher and births take place at an earlier maternal age (Monnier, 1991). Fertility has fallen to below-replacement level in several countries through the widespread usc of family limitation, most notably abortion, despite periodic government restrictions (Blayo, 1991).

Under state socialism, governments actively encouraged female

**Table 5.5** Total fertility rates, eastern Europe, 1990

| | |
|---|---|
| Estonia | 2.2 |
| Latvia | 2.0 |
| Lithuania | 2.0 |
| Bulgaria | 2.0 |
| Czechoslovakia | 2.0 |
| Hungary | 1.8 |
| Poland | 2.1 |
| Romania | 2.3 |

participation in the labour force, which was viewed as a quasi-obligatory role alongside motherhood (Heilig et al., 1990). The fulfilment of the latter role was encouraged, albeit unsuccessfully, by generous maternity benefits and leave arrangements and the establishment of extensive networks of childcare facilities.

The movement to democracy that has swept through former eastern bloc countries has been associated with the outright rejection of beliefs and policies associated with state socialism. A recent opinion poll in the newly independent state of Lithuania revealed the emergence of new conservative, anti-feminist attitudes (Vichnevsky et al., 1991). The consensus of these findings was that women should devote their lives to motherhood and their families; economic activity outside the home should be of secondary importance. Greater religious freedom has also contributed to an upsurge in pro-natalism. In Poland, laws banning abortion were not repealed, as in other eastern European countries, on religious grounds. In Bulgaria, Czechoslovakia, Hungary and the former Soviet Republics, anti-abortion campaign groups have emerged. Evidence of this sort suggests that there may be an increase in fertility in eastern Europe in the near future (Klinger, 1991). However, such a change in attitudes may not be sufficient to circumvent the fertility-reducing influences of liberal abortion laws and economic and social developments associated with the rapid transition to market economies (Vichnevsky et al., 1991). Recent Romanian statistics indicate 41,000 fewer births in the first nine months of 1991 compared with the same period in 1990. This shortfall has been attributed to the legalization of abortion and the emigration of citizens of reproductive age.

## Concluding Remarks

Low European fertility levels are a function of complex social, economic and political structures. Attempts to predict future trends, although essential for planning purposes, must be approached with caution. Couples do not, as economic theory predicts, always behave rationally when planning a family and it is therefore difficult to measure and forecast the way in which attitudes and values will change.

It seems unlikely, however, that there will be a significant reversal in fertility trends in the future. The baby boom generation is currently passing through the prime childbearing age-groups and yet no major recovery in fertility is to be seen. In France, West Germany and the UK a short-lived upturn in fertility was evident in the late 1970s which may be attributed to the echo effect as a large birth cohort passes through the reproductive age-groups. Elsewhere this was not sufficient to force overall fertility decline into reverse. Greater opportunities for female labour force participation and the fertility-reducing practice of cohabitation have counter-balanced even such a short-lived resurgence.

Improved facilities for the working mother may, as in the Swedish case, raise fertility back to replacement level. The eastern European experience has revealed that the relationship between fertility and childcare provision is by no means straightforward. The suggestion from some quarters that natural population growth may in the future be dependent upon the high fertility of certain pro-natalist ethnic minority groups has been discredited by recent research which has charted the diminution of majority–minority fertility differentials (for example, Kane, 1986; Schoorl, 1990; Sporton, 1991).

Geographical differences in fertility trends within Europe, once significant at the regional and later more apparent at the international level, are now being reduced. European social, economic and political unity should result in the gradual disappearance of both intra-Community and ultimately east–west differences in reproductive behaviour. Fertility levels in the New Europe are likely to remain, in the foreseeable future, the lowest in the world.

# 6

# Fertility Policies: a Limited Influence?

Jacqueline Hecht and Henri Leridon

Throughout Europe, there is no homogeneity in perceptions of and reactions to population policies. Fertility has dropped in almost all countries to unusually low levels, most often below replacement. However, the worry about demographic problems is far from general and there is no common wish to modify the curves of fertility trends. The current policies are fragmented, not always consistent and often uncertain if not hesitating. There are very few countries where the desire for a demographic renewal turns into concrete decisions and a well-developed policy.

Furthermore, explicit policies which are intended to influence demographic trends must be separated from implicit ones. For most observers, none of the European Community (EC) countries has an explicit population policy. Most countries display some reluctance towards population policies and even family policies, mainly in consequence of the remembrance of war and the pre-war period.

The measures which are taken – when they are taken – may still be direct or indirect, quantitative or qualitative. One may say that a population policy is voluntarist when its intention is to act directly on the demographic variables, and a population policy may be said to be derived when its intention is to exert an indirect influence upon these variables. The difficulty is to know which of these two patterns of action is the most efficient and to attempt to combine them in a single intervention (Tabah, n.d.). From this point of view, laws regulating contraception and abortion will not be considered here as part of fertility policies, since they are not intended, at least in developed societies, to change fertility trends, but rather they are regarded as contributions to basic individual liberties.

Voluntarist fertility policies emphasize two principal fields of action: the compensation for financial burdens mainly via allowances and tax concessions; and improvement of the family environment especially through the reconciliation of family and working responsibilities and assistance with the problems of childcare.

This means that policies intending to raise fertility are often part of social policies. Fertility policies almost always aim to reduce inequality: they ensure income redistribution and strengthen solidarity between the different age-groups and social classes. Social policies often precede family and fertility policies in western Europe; nevertheless these two kinds of policies are not absolutely interchangeable and identical in terms of their motives, their operation and their effects. They can result in contradictions between the various targets and, furthermore, some useful social and economic measures may unwittingly contribute to slacken nuptiality and to reduce natality. Economic and financial constraints make the rationalization of social and family policies inescapable. But the ceiling placed on allowances and the use of means-tested benefits make them appear more like devices with which to fight poverty than measures to offset family burdens.

This approach often reduces the family-oriented characteristics of policies which are often selective; they do not attempt to help 'all' families, but rather families 'in need', taking into account the family income, the number of children, the age of the beneficiaries and the status of the couple. These policies are therefore likely to favour low-income families, large families, young couples, old people and, because of the change in family structures, one-parent families. Many measures become means-tested, thus leading to a vertical redistribution of resources.

Fertility policies act mainly through social measures upon the mainspring of demographic dynamism: the family unit. The pre-eminent value of the family for legislators is widely recognized at the national level: eight European countries, among them France, Germany and Luxembourg, explicitly refer to the rights of families. Some countries have, or have had, a Ministry or a Secretary of State specifically in charge of family rights. These include France, Luxembourg, the Federal Republic of Germany, three Belgian communes and some German *Länder*; and in some, France for instance, family associations exert a powerful influence on official policy.

## Pro-natalist Policies in Western Europe

A fertility policy will therefore be a social and a family policy. But this policy, whether familial, social or demographic, must necessarily be an integral component of a more general policy. A social and family policy may be said to be pro-natalist if family allowances rise with the child's birth order since to give special privilege to parents with two, three or more children gives evidence of distinct demographic concern. Countries may even be classified according to the way family allowances are linked to birth order. If allowances are centred on one or more precise ranks, as is the case in France, it means that the demographic concern is already present, but if allowances rise with birth order, as is the case in Belgium, Luxembourg and the former Federal Republic of Germany, then one may speak of a strictly pro-natalist policy (Ekert-Jaffé, 1988).

Few western European states apply an explicit policy. For example, in Belgium the pro-natalist policy is much more implicit than active and in the former Federal Republic of Germany there is no explicit demographic policy, although certain pro-natalist components such as paid maternity leave and the increase of allowances according to birth order are present.

The only countries with an explicit family policy – France, Belgium, Luxembourg and the Federal Republic of Germany – demonstrated by the existence of government ministries, the use of political rhetoric and political labelling, do display pro-natalist tendencies, so one could conclude that explicit family policies represent at least partially a disguised population policy. It has been pointed out that none of the successive governments of the Federal Republic of Germany has pursued a population policy and even if demographic arguments are sometimes put forward in Luxembourg, for instance, family policy remains an element of social policy and no set of measures ever explicitly refers to the demographic state of the country.

But even though most EC countries have no concrete set of measures which are directed towards demographic targets, they have all developed social policies which may have some influence on the behaviour of families and by this means they will exert an influence on demographic variables.

## The Main Components of Current Policies

Four EC countries currently have a special interest in family policy: Belgium, France, Luxembourg and Germany, particularly the former German Democratic Republic.

The main axes upon which a social and family policy may work are the compensation for additional financial burdens by the payment of family allowances for children up to 16–25 years, which might increase for large families with low incomes, and intervention to alter general living conditions.

### *Family Allowances*

Some allowances do not depend on income, as is the case in France with birth and family allowances; others are paid in the form of income supplements to couples whose income does not rise beyond a fixed level. A third form of allowance is deliberately targeted to support a specific form of spending, such as that on housing, or to help families facing special difficulties like disability, single parenthood and so forth. In some countries allowances are paid without reference to birth order, but in most – Belgium, Luxembourg and Germany, for example – allowances do increase according to birth order. Young couples are also offered loans in Luxembourg and Germany.

### *Taxation Measures Favouring Families*

Taxation measures are numerous and highly diverse. First, family allowances are usually tax free. Second, there may be tax relief according to the number of children and for an unemployed spouse. The most efficient method of relief, exemplified by the French system of 'family quotient', is to reduce the progressiveness of taxation as income is increased by the addition of the incomes of the two spouses. This is achieved by first dividing the total income, or some part of it, by a ratio which is determined by the composition of the family and then multiplying the resulting tax liability by the same factor. A largely similar system of 'splitting' has been applied in Luxembourg and the Federal Republic of Germany. In addition, special provisions may be made for single-parent families.

### *Maternity and Paternity Leave*

Maternity leave lasts from 8 to 14 weeks in Belgium, 16 to 24 weeks in France, 16 weeks in Luxembourg, 14 to 18 weeks in the Federal Republic of Germany and 26 weeks in the former German Democratic Republic. Wages can be paid up to 100 per cent as in the Federal Republic of Germany or 90 per cent as in France. In addition, 'parental education leave' is increasingly available to the mother or the father, sometimes under specific conditions, with guaranteed re-employment at the end of the leave, but often there is no financial support. Even then the time spent away from work rearing children may be taken into account when calculating pension rights.

### *Family Amenities: Creches and Kindergarten*

There are wide variations in the provision of these amenities among the EC countries. For example, in the Federal Republic of Germany only 2 per cent of children under 3 were cared for in creches compared with 85 per cent in the German Democratic Republic. But for children aged over 3 the kindergarten system does meet needs more fully in the Federal Republic of Germany with 60 per cent of all 3–4 year olds and 90 per cent of 5 year olds finding places in nursery schools. In general, the supply of nursery school places is low compared with demand with the result that local communities are playing an increasingly important role.

### *Housing, Public Transport and Training*

On housing, public transport and training, special advantages are generally granted to large families with low incomes in France, Belgium and Luxembourg, but these measures may be regarded as part of social rather than pro-natalist policy.

### *Examples from Other European Countries*

In Europe, there are only a few instances of explicit population policies. Only Malta, the Netherlands and Turkey have set explicit targets aimed at achieving stationary or declining fertility. Cyprus intends to raise its fertility rate and some other countries, like Greece, Italy and Spain, have recently expressed some concern about the decline of the birth rate, but their family policies are only implicit. Italy has no formal family policy and no official institution taking direct care of families. But most political parties have begun to express concern over the declining birth rate and to advocate stronger support for families especially in the form of reduced taxation, longer maternity leave and the grant of child allowances up to the age of 3. Cash measures are more popular than services in kind; family allowances have become rather low and are means-tested.

In Spain, the 1978 constitution proclaimed in a rather general and abstract way that the state had a duty to protect children and families. A new family policy has been taking shape since 1975, but this should not be taken to mean that an explicit family policy will be created. The political leaders fear that current demographic trends might ultimately prevent economic growth and they are beginning to consider measures to increase the birth rate or at least to prevent its further fall.

In Portugal, a Family Ministry was created in 1980. Family allowances increase from the third child for poor families, but in general the progressive structure of family allowances according to birth order was abolished in 1985. The Portuguese state pays a wedding benefit, a birth allowance, a breast-feeding benefit and a specific means-tested allowance to needy families.

In the UK, 20 years ago little attention was paid to the social and economic problems of families. But since the early 1990s many political leaders have begun to speak about the importance of the family. Nevertheless, the UK still lacks an explicit family policy. The Irish Constitution does recognize the significance of the family and of marriage, although here too there is no explicit family policy.

In the Netherlands there appears to be no desire for an active population or family policy and the measures taken to favour families are not intended to have a demographic impact but to promote equality of opportunity. In its 1977 report the Royal Commission on Population urged the government to ensure a stationary population and established the Inter-departmental Committee on Population Policy to implement the matter. But while demographers plead for an active pro-natalist policy through indirect measures, feminist groups are opposed to such a policy.

### *The Case of the Nordic Countries*

The Nordic countries have always had a social policy which has often been viewed as a family policy, and in some cases even as a fertility policy.

There is no family policy in Denmark, but a policy centred on childhood. The Finnish government also wishes to improve the situation of families with children by developing the child benefit system and by raising the age limit. Since 1988 Finnish parents who have small children may take advantage of a six-hour working day as well as periods of leave for up to four consecutive days to take care of a sick child.

In Norway there seems to have been more interest in the fertility rate. The report of a government-appointed Population Committee published in 1984 recommended that the living conditions of families with young children should be improved and that fixed child benefits should be paid.

The total fertility rate in Sweden has recently reached replacement level at 2.2 children. While there is no explicit fertility policy there are social and family policies based on day care for children and parental leave. Family policy has always been an essential component of the Swedish welfare system. At the moment this allows women both to be economically active and to have a fertility higher than women in many other European countries (Nasman, 1991). A general allowance for dependent children is paid at a uniform rate for every child up to the age of 16, or 20 for those still in full-time education, but families of three or more children receive a supplementary benefit which adds as much as another half allowance for the third child and one and a half extra on top of the basic allowance for each additional child. Single parents receive a maintenance allowance or a maintenance subsidy that the non-custodial parent is obliged to pay. None of these benefits is means-tested or taxable. The pregnancy leave entitles a woman to 50 days of allowance within the last 60 days before the anticipated delivery date. And the birth leave offers 540 days of leave and 450 days of parental insurance. Of these 450 days, 360 are compensated by receipt of 90 per cent of normal wages. Since 1985, day care must be made available to every child between 18 months and six years. In 1988, 69 per cent of children benefited from this facility. The fees required from parents vary depending upon income, the number of children attending and the length of their stay.

### Measuring the Impact of Pro-natalist Policies

In the past 30 years enormous efforts have been expended on measuring the effectiveness of family planning programmes (see the reviews in United Nations, 1979, 1985a; Lloyd and Ross, 1987), much less attention has been paid to methodological issues in the evaluation of pro-natalist policies. In principle, we could regard the two problems as perfectly symmetrical: in one case, coercive measures are taken to discourage people from having children, and we can see how much fertility has been affected by this policy; in the other case, measures are taken to encourage people to have more children and the expectation is that fertility will rise, or at least it will cease to decline. But when we begin to examine in detail the techniques used in the evaluation of family planning programmes, it appears that some concepts are quite specific.

In the context of a family planning programme, a central

concept is one of acceptors. The programme is usually offered in specific areas, because its implementation requires specific and material means of action (clinics for consultation, channels for delivering contraceptives etc.); it is possible to know exactly where the programme is really operative and how many women or men have 'accepted' to enter the programme, and it is even possible to keep a record of these couples and to carry out follow-up studies of their reproductive behaviour. The corresponding concept for a pro-natalist policy would be the convinced beneficiaries, those who are eligible for the benefits offered by the law and who have been convinced to attempt conception in the expectation of these returns. Unfortunately, there is almost no way to identify such couples: first, because the policy is most often applicable at the national level, hindering any regional or sub-regional analysis; second, because couples are not asked to register when they find the pro-natalist law attractive enough to want one more child.

Another key concept for the measurement of the effectiveness of a family planning programme is the one of non-programme users of contraception. This category is important because some couples can switch from this group to the one of acceptors, increasing the gross impact of the programme but not its net effect. Such a substitution might be quite important: it is known that the first acceptors of a programme are those who would otherwise have used some kind of family planning method. In the context of a pro-natalist policy, similar substituting effects can be foreseen: in that case, we can expect a change in the tempo of fertility for those couples who were not far from deciding to have a child and take the opportunity offered by the new policy. But, here again, there is no way to identify these 'waiting-for-an-opportunity' couples.

One straightforward conclusion to be drawn is that the techniques of evaluation using the above mentioned concepts are not applicable to pro-natalist policies, which rules out about half of the methods available. Not only are methods relying on the number of births averted excluded, but also those based on matching techniques between acceptors and non-acceptors, on an individual or group basis.

Natural fertility, defined as the level of fertility in the absence of family planning, all other things being set equal, gives a useful reference to which the actual level of fertility can be compared. The difference between the two is a measure of the net impact of the programme. We cannot think of any corresponding concept for pro-natalist policy. Natural fertility is an estimate of maximum fertility, albeit under certain specified conditions, but the minimum theoretical level of fertility is zero, although it is of course known that in societies without any policy aiming at encouraging couples to have children, couples still do have some children.

We are left, finally, with three categories of methods that can be used in both contexts: first, comparing the actual (annual) level of fertility to an extrapolated (and thus hypothetical) trend; second, comparing the trends in fertility between socio-demographic groups (or between countries) that have been unequally exposed to the new policy; third, modelling the behaviour of couples, mainly by means of econometric models, to derive an estimate of the theoretical level of fertility in the absence of monetary incentives to have children.

The remainder of this chapter will be taken up with a report on the results of studies based on these various approaches and a brief discussion of an additional method which uses answers to survey questions on the possible changes of behaviour that could be generated by specific pro-natalist policies.

## Trend Analysis

The simplest, and most useful, way to judge the efficiency of a new policy is to compare the actual evolution of some index of fertility with an estimate of the expected trend without any change of policy. Most often the reference curve is determined by a mere extrapolation of the past trend; the extrapolation may even remain implicit. For instance, fertility was in decline before the new law was passed and the decline stopped thereafter; this discontinuity is seen as proof of the effect of the policy.

Festy (1986) has identified two main objections to the adoption of this procedure. The first problem results from the fact that in many cases, mainly in eastern European countries, the new policy included not only pro-natalist incentives but also restrictions on abortion. The latter cannot be seen as a mere disincentive, but as a plain constraint, and ideally its effect should be separated from the other changes, but in practice this is rarely feasible. The second objection deals with the problem of change in the tempo of fertility versus change in its magnitude. A typical example is provided by the German Democratic Republic. There the total fertility rate (TFR) rose from 1.54 in 1974 to 1.94 in 1980 (+26 per cent) and the change seems to have been highly related to the new policy adopted in 1976. In one of the most recent analyses of this experience, Monnier (1989) has compared the mean number of births by cohort as observed in 1984 with an expected number calculated by extrapolating the number of births reached in 1976, in the same cohort, on the basis of a regression between mean parity at age $x$ and mean parities at ages $x - 1$ and

$x - 2$. The parameters of the regressions were estimated for the period 1955–75. It appears that the policy effects might have been almost nil for generations over 25 years of age in 1976, but it might reach 0.2 children per woman for women aged 21–2 in 1976, so increasing the mean number of births by 13 per cent. This study supports others which suggest that the most significant effect was on second-order births and, but to a lesser extent, third-order ones.

Büttner and Lutz (1990) have used a different approach to deal with the same problem. They computed a period effect net of cohort and age effects by means of a model in which each annual fertility rate was expressed as a product of three components – age, period and cohort (hence the APC model) – plus an additional random component. The period effect is assumed to be specific to 1977–87 and was found to be almost constant and equal to 15 per cent for TFR in the age-group 18–30 over those years. This finding seems to be more positive than Monnier's, but actually the APC model does not deal properly with the changing tempo of fertility. One could conclude from these two studies that the impact of the policy developed in 1976 might have been at least 0.2 children per woman (+13 per cent), a value already reached at age 30 in the most responsive cohorts, those born round 1955. However, this will probably not be sufficient to bring the final number of births in these cohorts up to the level of replacement fertility (a TFR of 2.1).

### Comparing Countries or Socio-economic Groups

Let us continue with the example of the German Democratic Republic and the law passed in 1976 which led to divergence in the fertility trends of the two Germanies. During the years 1959–74, the TFRs of both East and West Germany were not only highly correlated but almost identical, increasing from 1959 to 1965 and declining thereafter. It is therefore tempting to consider the difference that appears after 1976 as being purely attributable to the policy developed in the East, and indeed the TFR of East Germany has constantly exceeded the West German rate by 0.4–0.5 children between 1977 and 1985 (although the difference has been reduced in more recent years). But here again one needs to remove the effect of changes in the tempo of fertility from this trend in order to evaluate the pure effect of its more durable component.

The Second World War has generated a strong discontinuity in the fertility trend of many Western countries. In France this also corresponds to the implementation of significant pro-natalist and pro-familial laws generally known as the *Code de la Famille* (1939). Chesnais (1985) has compared the fertility levels of France before and after the war with the mean fertility of a group of other countries over the same period. Just before the Second World War, French fertility was below average by 0.2 children per woman; it exceeded the average by about 0.4 children in the period 1946–55; thereafter the difference was reduced and even cancelled by 1975. However, these comparisons are only based on period measures; they should certainly be qualified in terms of cohort fertility measures. Chesnais has also pointed out that changes in the tempo of fertility might be useful from a policy perspective if they result in an increase in the annual number of births at a time when such a rise is 'good for the age-pyramid'.

In fact the first family allowances appeared in France in 1919, but they were only made available to civil servants. For employed persons in the private sector similar allowances were introduced progressively between 1920 and 1930. They became general in 1939 with the introduction of the *Code de la Famille*. For self-employed persons, however, the allowances remained very low until the mid-1950s. Calot and Deville (1971) exploited these lags between social groups in the implementation of family policy to estimate the specific effect of the policy on the general increase in fertility after the Second World War. They concluded that out of a general increase of 18 per cent almost half (8 per cent or 0.16 children per woman) could have resulted from the new policy introduced in 1939 and maintained during and after the war. This may be a unique example of fertility differentials within a single country that can be analysed in relation to different policies and estimated in terms of cohort fertility.

### Modelling Fertility Behaviour

Fertility studies usually take into account a series of socio-economic indicators either for multivariate analysis or for inclusion in behavioural models. When the scale of analysis is national, some indicator of the intensity of pro-natalist policy in each country may be included. This has been done by Ekert-Jaffé (1986) in a regression analysis which incorporated the level of female wages, the female employment rate and the intensity of pro-natalist policy as exogenous variables to estimate TFRs in seven western European countries in 1971–3. The variable measuring average female wages proved the most significant with the pro-natalist variable coming second. With a policy like the French one, fertility could be increased by 0.2 children per woman and Ekert-Jaffé concludes that if the total cost of all children were to be compensated by specific allowances then the effect on the TFR might reach 0.5 children.

In association with A. Blum, Blanchet (1987) has developed a micro-economic model inspired by the so-called 'new home economics' approach. For each birth order, the decision to have a child is said to depend on the cost of the child and the amount of resources the parents are ready to spend to raise all their children. The costs are direct (food, clothing etc.) and indirect (childcare if both parents are working, for instance) and there are opportunity costs if one parent stops working to take care of the child. Public allowances will reduce the costs or the burden of the families. The input variables are the parity progression ratios and the labour force participation rates for women by parity. Based mainly on French experience, Blanchet's analysis allows the impact on fertility of parity-specific measures to be computed, as well as the estimation of their economic efficiency. For example, it shows that a full compensation of the average cost of the first child would increase the TFR by only 0.1, whereas compensating for the average cost of the third child would increase fertility by 0.3. With an allowance equal to half the cost the impact would also be reduced by half. Although the model is not intended to be additive for the various ranks, it does confirm that the total impact of the French policy would be about 0.2–0.3 children per woman, and that a full compensation for the cost of all children would increase fertility by less than one child. In addition, it shows that the economic efficiency of measures applying to parities one and two is rather low.

### Survey Questions on Intentions

One simple way to forecast the effectiveness of pro-natalist legislation would be to ask the people themselves about their reactions to such measures. For instance, in a survey undertaken in Quebec, women were asked about the measures they thought would be most effective in helping parents to raise their children and then, selecting the single most effective one, whether its implementation could make them want to have more children. Obviously, there is no opportunity to check whether the answers really do have a predictive value since there are no examples of governments following such questioning by actually implementing the suggested policy. But there are some positive indications that this form of questioning is of value especially in certain circumstances. For example, the proportion of French women reporting that they would consider having an abortion in the case of an unwanted pregnancy was probably close to reality (Leridon et al., 1987,

p. 255). Moreover, the Quebec survey mentioned above also revealed that the women interviewed did not claim unrealistic intentions. Only 10 per cent of married women aged under 35 answered 'yes, they probably would want more children under specific conditions'. All in all, the total impact on fertility if all intentions were translated into action would be less than 8 per cent, or 0.2 children per woman (Henripin and Lapierre-Adamcyk, 1974). Again, this is an estimate of the same order of magnitude as that found using very different methods.

### Concluding Remarks

The experiences of several eastern European countries, apart from East Germany, have also been examined in some of the publications already referred to. In all cases the results were much less convincing: at most, fertility reacted with a transitory increase followed by a rapid return to the pattern of fertility current before the introduction of the pro-natalist policy. Even more drastic measures, such as the denial of access to abortion, do not seem to have had long-term effects. Reviewing a quarter-century of the experience of induced abortion in eastern Europe, Thomas Frejka (1983) concluded:

> Restrictions in the liberal abortion legislation – which were almost always coupled with pro-natalist economic and social measures – generated some fertility increases of a transitory nature. In the long-run, however, it would seem that even though the incidence of induced abortion might be high, fertility levels have been only marginally affected by abortion legislation.

We must remember that even a temporary rise in the number of births induced by changes in the tempo of fertility might be useful. Let us take the extreme case of Romania where the dramatic rise of fertility, amounting to a doubling of the annual number of births produced by the sudden restrictions placed on abortions at the end of 1966, severely distorted the age-pyramid and had negative effects on perinatal and infant mortality. From 1970 to 1979 it stabilized the annual number of births at a level close to that of 1949–57, well above the minimum reached in 1965–6 and below the exceptional values of 1967–8.

The Romanian story is a very sad but instructive example of how dictatorial regimes and coercive measures may be more efficient than the incentive-oriented policies employed in democratic societies. The studies considered in this chapter tend to support the view that at maximum a 10–15 per cent increase in fertility can be expected as the result of the introduction of a strong pro-natalist policy. And even this assumes that the total cost of children will never be fully compensated for, and the policy is likely to require periodic reintroduction or reinforcement to remain effective.

On a more general level this chapter has also shown that fiscal and social legislation can take into account the cost of children in many different ways. Will it be possible to harmonize European family policies? A common European policy should result in a fair and equal treatment of families because, as Laroque (1987) has remarked, 'a great single market without children and soon without men' is difficult to imagine. It will be necessary to try and co-ordinate the systems of social security and regulation of the flows of migrants within the Community. With respect to family policies proper, the main problem is likely to be the management of time both at home and at work; part-time working, paid parental leave and the impact on the family of social policies all have to be considered further.

Of course, the social policy of the Community depends on its economic targets, and its family policy does not appear to exist as a separate entity even now. Nevertheless, the EC is becoming aware of the need for such a specific form of policy, and in the particular area of migrant workers and their families measures were even implemented as long ago as 1964. Clearly, the most advanced pro-natalist or even pro-family policies developed in some countries will not be imposed on the others, but a minimum degree of harmonization will be necessary in the new environment of free movement of goods, capital and, above all, people.

# 7

# Age and Sex Structures

Pierre-Jean Thumerelle

The population of the European Community (EC) has one of the world's oldest age structures. Never has the contrast between the ageing European population and the youthful population of the rest of the world been so great. The EC population represents 6.4 per cent of that of the world as a whole; it contains 3.6 per cent of the under 15 year olds but 15 per cent of those aged 65 and over. These figures speak for themselves: the strengthening of the EC's economic potential is counter-balanced by the weakening of its human capital.

This situation should not be surprising. According to demographic transition theory, populations begin to show signs of ageing from the point when they reach the end of transition and growth rates stabilize. For a long time the loss of population was characteristic of rural areas, but it has now become a phenomenon of urban-industrial societies. In just a few decades demographic ageing has affected the whole of Europe, including the east and the old Soviet Union.

Although ageing is the dominant characteristic, we should not be tempted to limit the geographical study of age structure merely to the phenomenon of ageing. The increase in life expectancy among adults, the renewal of generations and more generally the relationships between major age-groups are very unequal from country to country and region to region. These differences have a particularly important bearing while Europe's demographic dynamism is weak, and the inertia accumulated in the age structures strongly influences the future vitality of populations.

## Many Old; Few Young

The age-pyramid of the EC population in 1990 is a perfect illustration of population ageing, with a markedly narrow base and an expanded adult age-group (figure 7.1). The age-groups 0–4 and 5–9 appear almost identical, but in reality this is not the case. Although the base of the pyramid has recently ceased what had seemed to be a never-ending narrowing, natality has not recovered sufficiently to establish a lasting reversal. Rather, it reflects the fact that the children of the post-war baby boom have themselves reached the reproductive ages. Each new generation is markedly smaller than the preceding one. On the other hand, between the ages of 25 and 40, where losses due to mortality remain small, the steps of the pyramid demonstrate the force of the post-1945 demographic revival, the effects of which lasted right up until the end of the 1960s. The relative importance of the older age-groups is proof of both the ageing of the population in general and the low level of mortality up to and including middle age.

Since the Community is a relatively recent political construct it may be unreasonable to contemplate the existence of a single European age structure; rather there exist a variety of structures reflecting national and regional historical processes, divergent socio-economic contexts and different mortality patterns (see chapter 4). In order to study these differences and their effects on the age structure of the EC population we must be able to compare the numbers remaining from 100 successive generations in the 12 member states and in their major geographical components. This task is only possible if one selects a small number of key indicators that can be used for summary and comparison. Figure 7.2 shows two such indices for 1990: the median age and

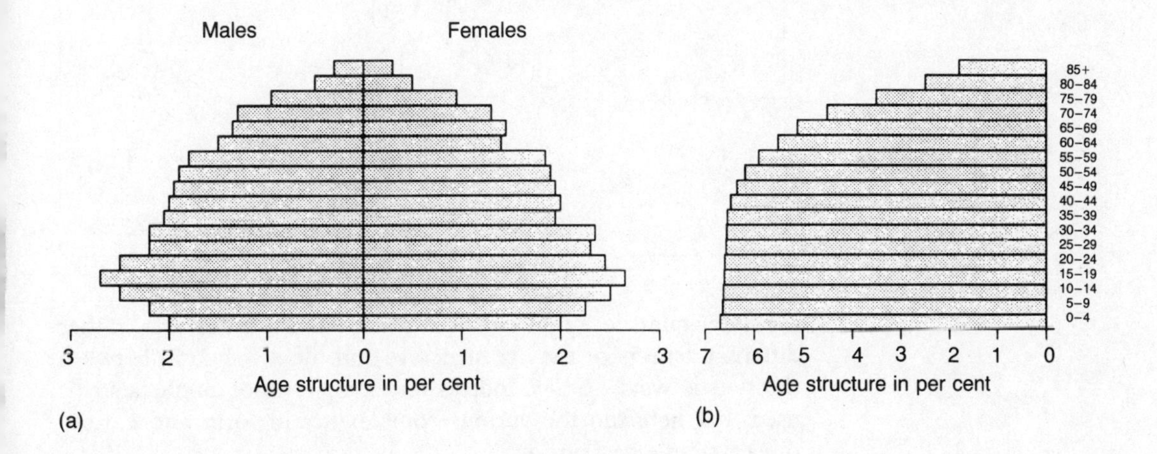


**Figure 7.1** Age structure of (a) the European Community population in 1990 and (b) a stationary population with an equivalent mortality level.

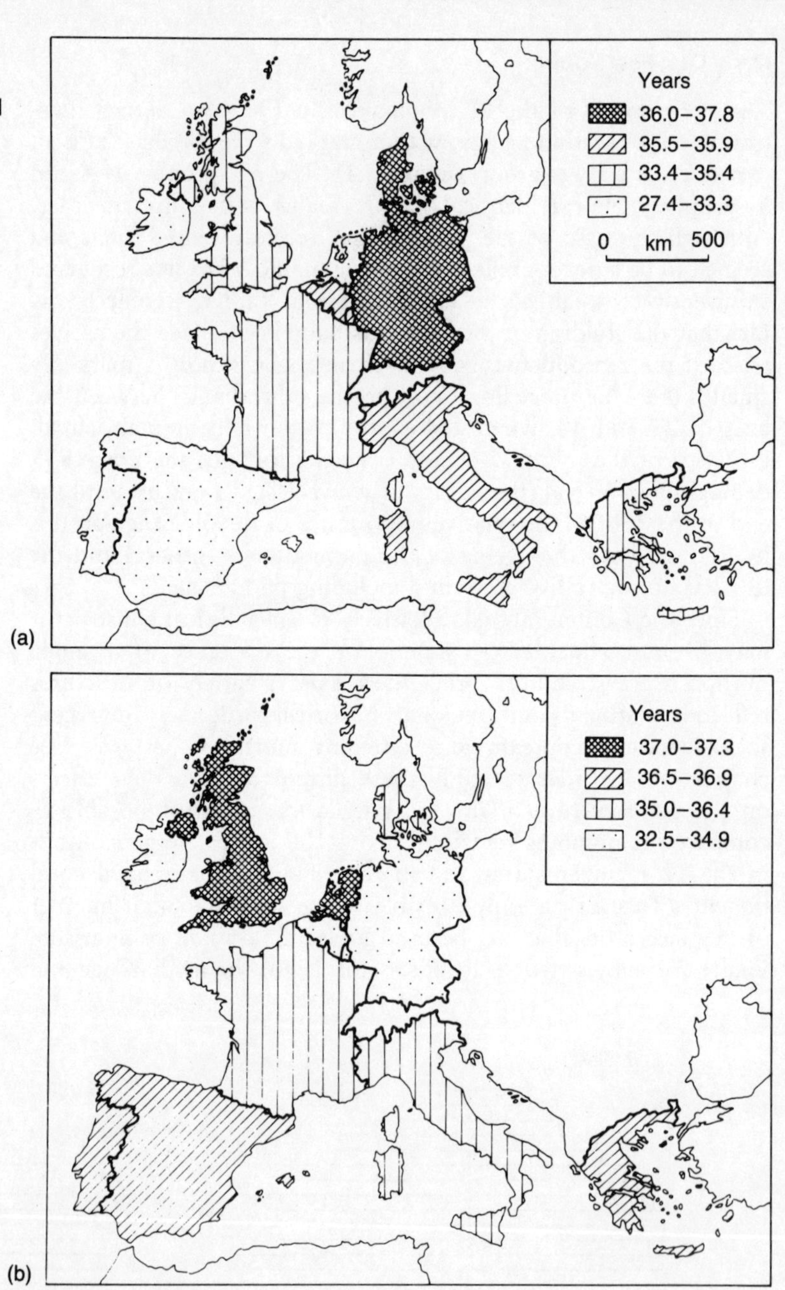


**Figure 7.2** National variations (1990) in (a) median age (years) and (b) inter-quartile age range (years).

the inter-quartile range or midspread. Each picks out rather different aspects of the age structure and the association between the two is weak. Other indices and simple ratios might also be used, but here too the various complexities of form and pattern thwart simple description.

An alternative approach might employ a standard age structure against which comparisons could be made. For this purpose I have selected a stationary model that reflects the prevailing European mortality level and which is by definition unaffected by crises and cyclical movements which influence real populations (also shown in figure 7.1). Deviations from this standard may be further classified into age structure types. Seven of these are shown in figure 7.3.

The first type, represented by Hamburg, shows an acutely aged population with excess in the 65 and over category balanced by deficit in the 15 and under category. Much of north, central and south-west Germany also matches this type. The second type, illustrated by Tuscany, is rather less accentuated at the extreme ages. It is to be found especially in Germany and northern Italy. The third type is even more muted. The birth rate has remained sufficiently high long enough for the deficit in the younger age-groups to appear less marked, but still there is an excess in the retirement category. Wallonia, Denmark, much of Great Britain and southern France illustrate the third type. The fourth type matches the stationary population rather well, apart from the 65 and over age-group. Large parts of northern France and eastern Spain exemplify type 5 structures with a small deficit of those in the middle age-groups and surpluses of the very young and old. Type 6, which applies to large parts of southern Europe, lacks excessive demographic ageing. Finally type 7, exemplified by Pas de Calais and Eire, shows a relatively youthful age structure by modern European standards.

### Age Structures of the Future

Hardly any part of Europe can be assured that its population will replace itself in the near future. However, immigration from Third World countries and the possibility of more and more people emigrating from the former eastern bloc countries might compensate for the persistently low birth rate. But many countries are trying to avoid this happening.

According to the latest EUROSTAT projections based on current trends, the population of the EC will be little larger in the year 2000 than it is today (with only a 2 per cent increase). This would produce a population of at maximum 352 million inhabitants. From the turn of the century the size of the population would start to decrease, slowly at first, but then increasingly rapidly. Internal variations would remain perceptible because, although the populations of the Netherlands, France, the UK and Spain would continue to increase, the populations of Eire, Portugal and Greece would enter a period of stagnation while the

**Figure 7.3**
Classification of regional age-group types.

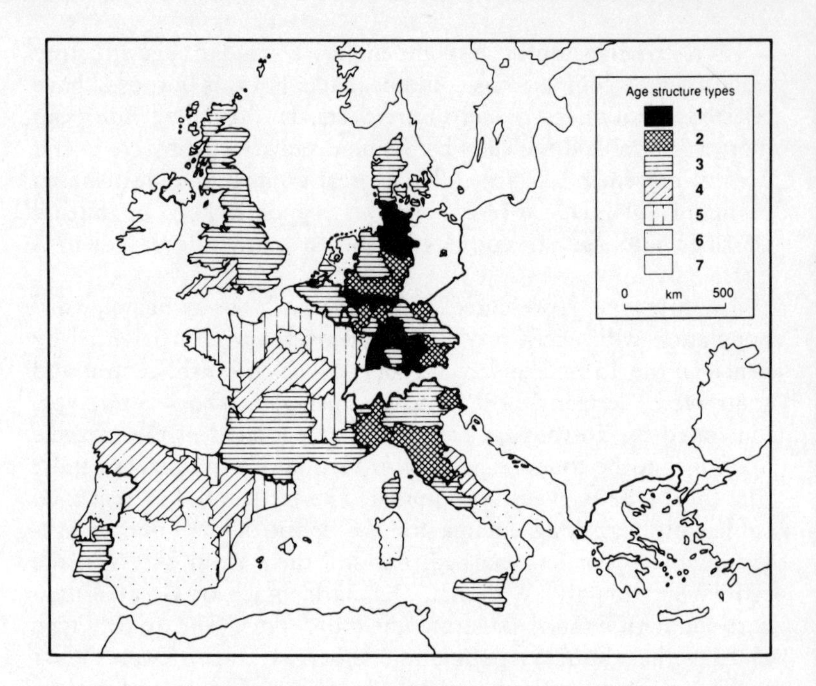


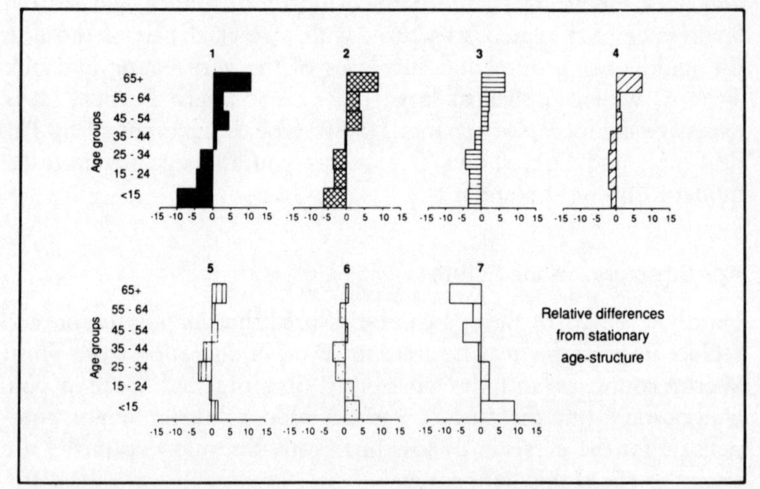


populations of such countries as Belgium and Germany would already be in decline.

Paradoxically, this would not inevitably result in an appreciable increase in age structure differentials. The silhouettes of the age-pyramids of the most aged countries are becoming more and more like each other and like that of the stationary structure used as the standard. During the twenty-first century the absolute size of the population could diminish while the relationship between age-groups stays the same; the age structure would be truly stationary and thus also stable. Youth would be a rare commodity. In Germany and Italy the under-15s would come to represent only 12 per cent of the population, or exactly half that of the over-64s. Only in Eire, the UK and Spain would the number of under-15s remain a fraction below that of the over-64s for a short time.

The relative and absolute importance of the elderly population would be felt everywhere in Europe. Except in Eire, the Netherlands and Portugal the over-64s would represent over 15 per cent of the population by the year 2000. By 2021 the percentage will be in excess of 20 in Germany, Belgium and Denmark and just under that figure in Italy, France and the Netherlands. By 1990 most Community countries had over 6 per cent of their populations aged over 74.

#### Concluding Remarks

In general, the size of the economically active population aged 15–64, which must bear the load of both young and old dependant populations, will vary little except in those countries where fertility has traditionally remained high, such as Eire and Spain. Here the increase will be noticeable, rising from 61 per cent in 1990 to perhaps 66 per cent in 2000 in both countries. The percentage of 15–64 year olds should be between 66 and 68 by 2000.

The immediate benefits that certain countries can draw from this situation will be felt for some while. For example, Germany and Italy, with 69 per cent in the economically active age-group, will bear only a slightly higher burden of aged dependants than France or the UK while bearing a smaller load of dependent young. The potentially active populations of the latter pair of countries are smaller than those of the former pair, although the differences will be reduced in about ten years' time.

Taking into account the inherent inertia of demographic structures, the relative advantage felt by countries with ageing populations will become a profound structural handicap in the twenty-first century and only a marked revival in reproduction is likely to alter the course of this particular age structure.

# 8

# Demographic Ageing: Trends and Policy Responses

Anthony M. Warnes

All readers will already have a view about population ageing, but it is important to begin this chapter by reviewing some of the concepts currently in use and focusing on the more important issues. First, we need to understand the mechanisms or 'proximate causes' of demographic ageing, not least to specify as well as we can the assumptions employed in population projection models. Second, we need to understand the implications for individuals and societies of age structure change. This is by no means a simple task as ageing occurs simultaneously with myriad economic and social transformations. Third, there are applied problems, beginning with the impacts of societal ageing on the functions, spending and taxation policies of governments and their agencies and extending to the dissemination of practical information to individuals about the changing circumstances of old age. There are links between the problem areas. For example, if we misunderstand ageing and misspecify projection models, then a misreading of the tasks for government becomes inevitable.

Neither a scientific interest to specify the mechanisms or underlying causes, nor a welfare interest to promote the standard of living, health and social integration of elderly people lies behind the intense recent interest in social ageing. This chapter offers some redress. It has three principal aims: to present a short account of the diversity of ageing trends and dynamics among the nations of Europe; to examine the contribution of late-age mortality change to survival and population ageing; and to examine critically the political responses to the phenomenon.

Commentaries on the rapid alterations in the age structure

of European national populations abound, at least in northern and western Europe (see chapter 7). They range from articles in popular magazines and newspapers to a proliferation of national and international policy analyses and position papers (Council of Europe, 1985, 1990a; Noin and Warnes, 1987; Coopmans et al., 1988; OECD, 1988a; Nijkamp et al., 1990; Dooghe, 1991; Walker et al., 1991). There certainly have been rapid and unprecedented absolute and relative increases in the older population during this century. From 1950 to 1990 the European population aged 65 and over virtually doubled, from 34 to 67 million, compared with a 35 per cent increase in the population aged 15–64 and a 2.2 per cent decrease in the number of children aged 0–14 (table 8.1). The elderly's share of the continental population has climbed from 8.7 per cent to 13.4 per cent, compared with a slight rise in the working age-group's share (from 65.9 to 67.0 per cent) and a fall in that of children from one-quarter to one-fifth (from 25.4 to 19.6 per cent) (table 8.2). In the European Community (EC) 12 of 1992, the elderly population had increased slightly more slowly over the same period from 24 to 46 million.

**Table 8.1** The increase in Europe's elderly population (65+ years), 1950–2010

| *Area*[a] | *1950 (1,000s)* | *Change*[b] *(%)* | *1970 (1,000s)* | *Change (%)* | *1990 (1,000s)* | *Change (%)* | *2010 (1,000s)* |
|---|---|---|---|---|---|---|---|
| 65+ years | | | | | | | |
| East | 6,195 | 73.1 | 10,725 | 19.7 | 12,834 | 26.7 | 16,256 |
| North | 7,465 | 36.9 | 10,218 | 27.1 | 12,988 | 6.0 | 13,769 |
| South | 8,067 | 57.5 | 12,706 | 44.5 | 18,355 | 33.9 | 24,582 |
| West | 12,376 | 53.3 | 18,970 | 19.1 | 22,597 | 24.3 | 28,083 |
| EC[c] | 24,415 | 48.8 | 36,320 | 27.8 | 46,406 | 21.9 | 56,591 |
| Europe[d] | 34,150 | 53.6 | 52,455 | 27.2 | 66,697 | 24.0 | 82,696 |
| 15–64 years | 258,673 | 18.5 | 292,644 | 14.0 | 333,486 | 2.0 | 340,028 |
| 0–14 years | 99,701 | 15.4 | 115,033 | −15.2 | 97,557 | −7.3 | 90,400 |
| All ages | 392,523 | 17.2 | 460,132 | 8.1 | 497,441 | 3.2 | 513,637 |

[a] East comprises Bulgaria, Czechoslovakia, GDR, Hungary, Poland and Romania. North comprises Denmark, Finland, Iceland, Ireland, Norway, Sweden and the UK. South comprises Albania, Greece, Italy, Malta, Portugal, Spain and Yugoslavia. West comprises Austria, Belgium, France, FRG, Luxembourg, Netherlands and Switzerland.
[b] The percentage changes are from the earlier (column to the left) to the later (column to the right) dates. 1990 and 2010 figures are 1985-base projections.
[c] The European Community in 1992 comprised Belgium, Denmark, Germany (the data refer to the pre-1991 Federal Republic), Greece, France, Ireland, Italy, Luxembourg, Netherlands, Portugal, Spain and the UK.
[d] The regional figures are based on weighted means of national figures, and some small territories are not included, e.g. Gibraltar and Liechtenstein. Consequently the sum of the regional figures does not equal the European total.
*Source*: United Nations population data and projections

**Table 8.2** Progress of the relative elderly (65+ years) population share, 1950–2010

| | *Elderly population (%)* | | | | *Change in elderly population (%)* | | |
|---|---|---|---|---|---|---|---|
| | *1950* | *1970* | *1990* | *2010* | *1950–70* | *1970–90* | *1990–2010* |
| 65+ years | | | | | | | |
| East | 7.0 | 10.4 | 11.3 | 13.5 | 2.0 | 0.4 | 0.9 |
| North | 10.3 | 12.7 | 15.5 | 16.1 | 1.1 | 1.0 | 0.2 |
| South | 7.4 | 9.9 | 12.7 | 16.3 | 1.5 | 1.3 | 1.3 |
| West | 10.1 | 12.8 | 14.5 | 17.9 | 1.2 | 0.6 | 1.1 |
| Europe | 8.7 | 11.4 | 13.4 | 16.1 | +1.4 | +0.8 | +0.9 |
| 15–64 years | 65.9 | 63.6 | 67.0 | 66.2 | −0.2 | +0.3 | −0.1 |
| 0–14 years | 25.4 | 25.0 | 19.6 | 17.6 | −0.1 | −1.2 | −0.5 |
| All ages | 100.0 | 100.0 | 100.0 | 100.0 | | | |

The percentage change represents the annual compound rate of change over 20 years of the share of the total population in the age-group.

That the increase in the relative elderly share is primarily the result of a falling birth rate is widely understood. Table 8.2 amply demonstrates the impact of smaller families on the European age structure, particularly since 1970. Fuller historical and regional accounts of the progress of ageing among European countries are available, although no decade by decade assessment is known of the contributions of reduced fertility, reduced mortality and migration exchanges (Laslett, 1985; Chesnais, 1986; Clarke, 1987; Warnes, 1989a; Tabah, 1991). But the perception of demographic ageing is rarely sophisticated and, at least in Britain, it has been strongly influenced by Anglo-American neo-liberal or new-right politics. As a result, the dominant representation of ageing has been as a macro-economic problem, especially in terms of the obstacle it represents to the reduction of public expenditure and taxation (Gillion, 1991). The perspective has spread widely and has supported reductions in state pension entitlements and in funding for social and health services targeted at old people, despite challenges from social analysts including gerontologists and those from the political left. Some critics of the alarmist and problematic views celebrate the reduction of premature deaths, the combating of diseases and disorders and the extension of life; others point to the continuing over-representation of poverty in old age, particularly among widowed women, as well as the amount that needs to be done to raise the housing standards and general well-being of elderly people.

### European Ageing in Perspective

Magazine articles can give the impression that societal ageing is something rather new, a characteristic of the last two decades or even a post-modern characteristic. In fact, the mean life expectancy at birth of human beings has probably been increasing in leaps and bounds for 100,000 years (Hendricks and Hendricks, 1986, ch. 2). Among privileged and aristocratic European women, several studies have indicated a steady increase from around the low 30s in 1600 to the low 50s by 1900. Undoubtedly, however, changes in survival and reproductive behaviour in the modern era have brought about unprecedented age structure changes. Very low rates of infant mortality and fertility are raising the elderly population from around one in 20 of the total population to one in five. This modern age structure transition began in France in the mid-nineteenth century and occurred more widely among European countries in the first decades of this century (Warnes, 1989a).

The immense diversity of the experience and current position of European countries is a healthy corrective to an understanding formed by the excellent, accessible but dominantly alarmist analyses of the different experience of the USA. In the USA, as in Australia where problematic fiscal perspectives also have great influence, special factors have been comparatively high fertility and in-migration during the first three-quarters of this century and the relatively late development of public support for elderly health care. The former has produced faster rates of growth of the absolute number of elderly people; the latter has induced very rapid increases in public expenditure on the elderly.

While the elderly share of the population is becoming increasingly similar in the USA and Europe, from around 9.5 per cent aged 65 and over in 1960 to 13 per cent in 1990 and to perhaps 18 per cent in 2020, the USA's total population will be increasing much faster than Europe's (18.5 per cent compared with 3.2 per cent during 1990–2020). Consequently, although Europe's ratio of the elderly to working age population is higher than in either the USA or Australia, the projected percentage increase in the elderly population to 2020 is lower at 43 per cent than in either the USA (63 per cent) or Australia (88 per cent). Between 1960 and 1985 the four largest European economies – France, Germany, Italy and the UK – increased public expenditure on old age pensions as a fraction of gross domestic product (GDP) by 85 per cent to 11.7 per cent, compared with 7.2 per cent in the USA, but the projections for the coming decades are more favourable (OECD, 1988b, table C.1a). Similarly, expenditure on public

**Table 8.3** The level and pace of demographic ageing by continents, 1970–2010

| | *Population 65+ years, 1990* | | *Percentage increase in* | *Percentage increase in* |
|---|---|---|---|---|
| | *1,000s* | *% of total* | *1970–90* | *1990–2010* |
| Africa | 19,426 | 3.0 | 73 | 89 |
| North America | 34,485 | 12.5 | 59 | 23 |
| Latin America | 21,061 | 4.7 | 89 | 82 |
| Asia | 155,424 | 5.0 | 84 | 79 |
| Oceania | 2,409 | 8.0 | 50 | 61 |
| Soviet Union | 27,647 | 9.6 | 54 | 43 |
| Europe | 66,697 | 13.4 | 27 | 24 |
| Less developed | 179,827 | 4.4 | 83 | 83 |
| More developed | 145,828 | 12.1 | 45 | 33 |
| World | 328,115 | 6.2 | 64 | 60 |

The figures for the aggregate regions incorporate estimates for certain small territories not included in the continental estimates.
*Source*: United Nations population data and projections

health care has been increasing much more slowly in Europe. During 1950–87, the four major European economies increased the share of GDP spent on this heading by 90 per cent to 5.9 per cent, compared with a 250 per cent increase to 3.5 per cent in the USA (OECD, 1987b, table 18; OECD, 1990, table 1).

Another salutary reminder of the longevity of the process of demographic ageing in Europe comes from comparisons with other continents. Europe's elderly population has increased by just over a quarter during the last two decades. Every other continent has experienced at least double this rate of growth, and throughout the less developed world the rate of change has been three times as great. The disparity is likely to widen over the next 20 years, with the increase in Europe and North America falling just below one-quarter and the elderly population of Africa, Asia and Latin America increasing by more than 80 per cent (table 8.3). Although Europe does have the most elderly population of all continents, given its comparative affluence, its strong welfare systems and the relatively modest anticipated growth, it is stubbornly Eurocentric to regard demographic ageing as simply a home-grown problem.

## The Progress and Mechanisms of Ageing in Europe

The age structure of a population is a complex record of the history of fertility, mortality and age- and sex-selective net migration over the previous 80 years or more. A rise in the mean age of a population can occur as a result of migration losses of young people, an effect that has been widespread in the more remote rural areas of Europe and which persists in the poorer problem regions of the south and east. It can also occur as a result of migration gains of elderly people. This phenomenon has been spreading quickly during the last quarter-century and particularly affects the coastal areas of England and Wales, France and some parts of Iberia and Greece (figure 7.3).

The usual reason for a persistent and substantial growth of the absolute number of elderly people is a recent history of high fertility, as characterized by Iberia, Italy, the Netherlands and Greece. The usual reason for a rise in the elderly population share is a sustained fall in fertility and this has characterized, in varying degrees, all parts of Europe. Relatively low fertility in France in the late eighteenth and early nineteenth centuries put it foremost in the ageing process. By 1850 more than 6.5 per cent of its population was aged 65 years and over, and by 1900 this had risen to just over 8 per cent. Many nations of northern and western Europe saw the first significant increases in their elderly share in the 1910s and 1920s. The change continued strongly until recently. The average annual increase in the elderly population of Great Britain exceeded 2 per cent throughout 1921–51, but it has since moderated to 0.6 per cent during the 1980s. And throughout Europe it was slower during 1970–90 than in the previous two decades, while various projections suggest that it will be slower still in the coming 20 years.

Another misconception is that within Europe ageing is strongest in the north west. In fact, the most rapid changes are now taking place in the south where fertility has recently fallen sharply from relatively high levels. During 1970–90 the elderly population of the southern quadrant of Europe grew more than twice as fast as in the rest of Europe (45 per cent compared with 21 per cent) (table 8.1). In the next 20 years it is expected that there will be a further one-third increase in the south compared with a 6 per cent increase in the north. In the next few decades, ageing will be most rapid in eastern Europe. By the second decade of the next century, its elderly population growth at 21 per cent will have overtaken that of the south with 12 per cent.

There are of course national differences within each broad region. Some still reflect losses of young adults during and immediately after the Second World War; in particular the elderly population of East Germany fell from 1970 to 1990. Before the transformations of 1989, the former Democratic Republic was expected to show a strong resurgence of ageing during 1990–2010, restoring its leading position in the eastern group. Its ageing prospects now depend upon the movement of working age people, that is, the balance of losses of its native working age population set against gains of ethnic Germans from Poland and the former Soviet Union. Elsewhere in eastern Europe, Poland, Romania and Bulgaria have both the least aged population and the fastest rates of recent and prospective growth (figure 8.1). Increases in the very old population – 80 and over – in this region will be relatively modest up to 2000 but much higher in the first quarter of the next century (figure 8.2).

The macro-region with the greatest contrasts is the north, for Sweden has the most aged population and Iceland and Eire among the youngest populations of Europe (table 8.4). Finland has shown the third fastest growth of the elderly population over the last 20 years and Eire the slowest. During 1990–2010, it is expected that Norway will experience a decrease and the UK and Sweden only a

**Figure 8.1** National projections of the population aged 65 and over, 1990–2010: (a) population in 1990; (b) projected increase, 1990–2010; (c) share of total population, 1990; (d) projected percentage increase, 1990–2010.

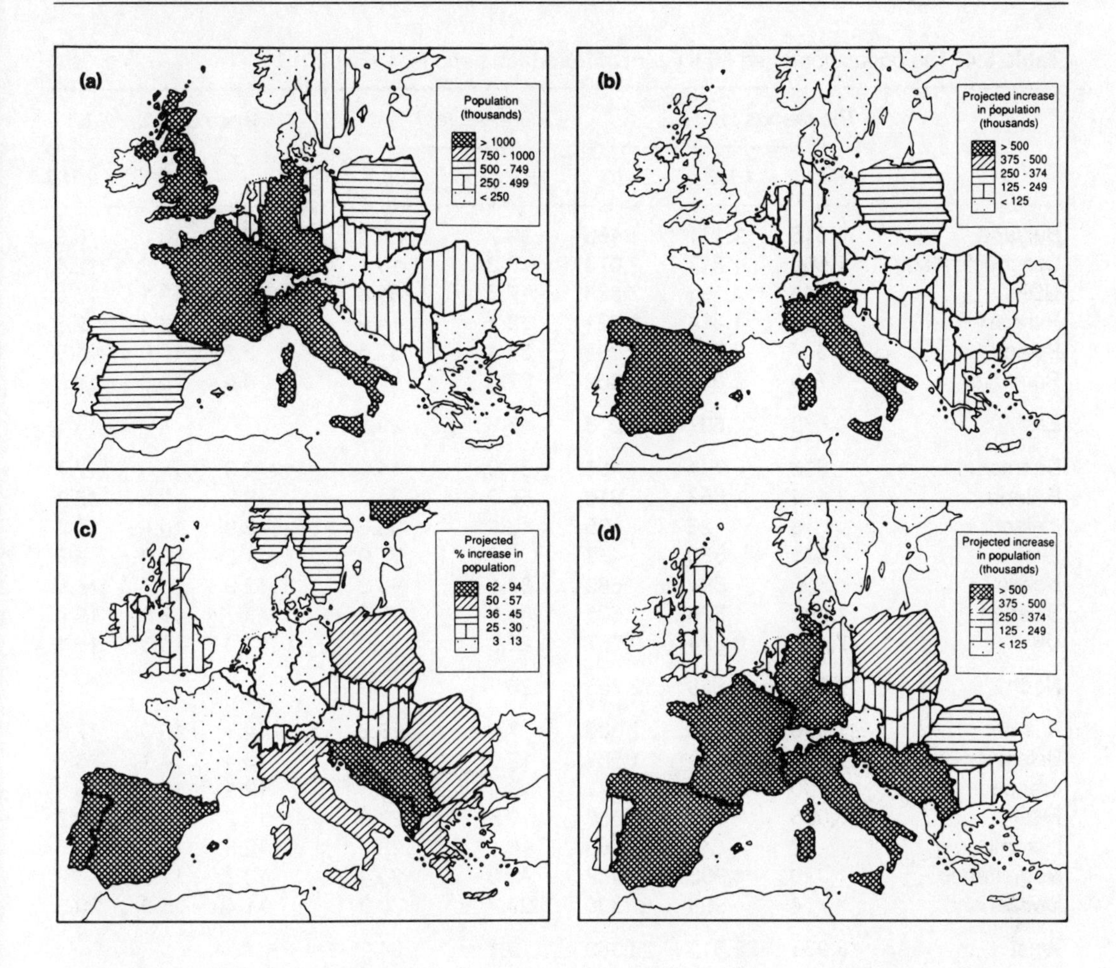


**Figure 8.2** National projections of the population aged 80 and over, 1980–2025: (a) population in 1980; (b) projected increase, 1980–2000; (c) projected percentage increase, 1980–2000; (d) projected increase, 2000–25.

5 per cent increase in their elderly populations: no other European nations can expect less than a 14 per cent increase. By 2010, Eire will probably have by far the youngest population, but its fertility did not fall swiftly until the 1980s. Throughout this region the number of very old has been increasing relatively quickly during 1980–2000, but growth during the first quarter of the next century is expected to be more moderate (figure 8.2).

In western Europe in 1990, the elderly share was relatively high in all countries at 2.9–15.4 per cent, but the underlying dynamic processes varied considerably. The Netherlands and Switzerland have recently experienced the fastest growth of their elderly populations (table 8.4). Projections made in the 1980s suggest that by 2010 the old West Germany will have the most aged population and the Netherlands and France will have the youngest populations. The prospective situation of France will reverse its position

**Table 8.4** The population aged 65+ years of European nations, 1970–2010

| | *Thousands* | | | *Increase (%)* | | *Share of total (%)* | | |
|---|---|---|---|---|---|---|---|---|
| | *1970* | *1990* | *2010* | *1970–90* | *1990–2010* | *1970* | *1990* | *2010* |
| Bulgaria | 815 | 1,171 | 1,468 | 43.7 | 25.4 | 9.6 | 13.0 | 16.2 |
| Czechoslovakia | 1,605 | 1,817 | 2,073 | 13.2 | 14.1 | 11.2 | 11.6 | 12.4 |
| GDR | 2,645 | 2,181 | 2,875 | −17.5 | 31.8 | 15.5 | 13.1 | 17.3 |
| Hungary | 1,191 | 1,414 | 1,621 | 18.7 | 14.6 | 11.5 | 13.4 | 15.5 |
| Poland | 2,667 | 3,557 | 4,894 | 33.4 | 37.5 | 8.2 | 10.0 | 11.5 |
| Romania | 1,750 | 2,397 | 3,302 | 37.0 | 37.7 | 8.6 | 10.3 | 13.2 |
| East | 10,673 | 12,537 | 16,233 | 17.5 | 29.5 | | | |
| Denmark | 606 | 794 | 906 | 31.0 | 14.1 | 12.3 | 15.5 | 17.7 |
| Finland | 424 | 657 | 816 | 55.0 | 24.2 | 9.2 | 13.2 | 15.9 |
| Iceland | 18 | 26 | 35 | 45.1 | 33.6 | 8.9 | 10.4 | 12.1 |
| Eire | 331 | 350 | 437 | 5.7 | 24.9 | 11.2 | 10.3 | 9.8 |
| Norway | 500 | 691 | 689 | 38.1 | −0.3 | 12.9 | 16.4 | 15.6 |
| Sweden | 1,101 | 1,526 | 1,605 | 38.6 | 5.2 | 13.7 | 18.3 | 19.4 |
| UK | 7,177 | 8,824 | 9,267 | 22.9 | 5.0 | 12.9 | 15.5 | 16.1 |
| North | 10,157 | 12,868 | 13,755 | 26.7 | 6.9 | | | |
| Austria | 1,050 | 1,124 | 1,306 | 7.0 | 16.2 | 14.1 | 15.0 | 17.8 |
| Belgium | 1,294 | 1,461 | 1,667 | 12.9 | 14.1 | 13.4 | 14.7 | 16.6 |
| France | 6,550 | 7,752 | 9,271 | 18.4 | 19.6 | 12.9 | 13.8 | 15.6 |
| FRG | 8,006 | 9,323 | 11,987 | 16.5 | 28.6 | 13.2 | 15.4 | 20.7 |
| Luxembourg | 42 | 49 | 63 | 16.1 | 28.9 | 12.5 | 13.4 | 17.5 |
| Netherlands | 1,329 | 1,903 | 2,482 | 43.2 | 30.4 | 10.2 | 12.9 | 16.2 |
| Switzerland | 714 | 998 | 1,306 | 39.7 | 30.9 | 11.4 | 15.3 | 20.3 |
| West | 18,985 | 22,610 | 28,082 | 19.1 | 24.2 | | | |
| Greece | 976 | 1,376 | 1,886 | 41.0 | 37.1 | 11.1 | 13.7 | 18.4 |
| Italy | 5,867 | 8,140 | 10,541 | 38.7 | 29.5 | 10.9 | 14.2 | 18.4 |
| Malta | 29 | 36 | 47 | 22.8 | 31.8 | 9.0 | 10.2 | 12.5 |
| Portugal | 832 | 1,327 | 1,600 | 59.5 | 20.6 | 9.2 | 12.9 | 14.8 |
| Spain | 3,310 | 5,113 | 6,484 | 54.5 | 26.8 | 9.8 | 13.0 | 15.5 |
| Yugoslavia | 1,589 | 2,170 | 3,744 | 36.6 | 72.5 | 7.8 | 9.1 | 14.5 |
| South | 12,603 | 18,162 | 24,302 | 44.1 | 33.8 | | | |

of nearly two centuries standing. This macro-region is likely to have relatively rapid growth of its very old population during the first quarter of the next century.

In the south macro-region, Portugal and Spain have recently led all Europe in the growth of their elderly population, but over the next 20 years it was anticipated that the former state of Yugoslavia

**Table 8.5** Projected decrease of the elderly and the total population among OECD member nations in Europe, 2040–50

| | *65+ years* | | *All ages* | |
|---|---|---|---|---|
| | *2040 (millions)* | *2040–50 (% loss)* | *2040 (millions)* | *2040–50 (% loss)* |
| Austria | 1.72 | 12.79 | 7.18 | 3.62 |
| Belgium | 1.97 | 8.63 | 9.02 | 3.88 |
| Denmark | 0.97 | 11.34 | 3.94 | 6.35 |
| Eire | 0.75 | −10.67 | 4.44 | 0.45 |
| Finland | 1.02 | 5.88 | 4.42 | 4.30 |
| France | 12.74 | 4.24 | 56.07 | 2.51 |
| Germany (FRG) | 12.50 | 18.08 | 45.30 | 7.64 |
| Greece | 2.02 | 2.97 | 9.64 | 3.53 |
| Iceland | 0.06 | 0.00 | 0.29 | 0.00 |
| Italy | 11.67 | 11.74 | 48.34 | 5.73 |
| Luxembourg | 0.08 | 12.50 | 0.37 | 0.00 |
| Netherlands | 3.43 | 13.12 | 12.85 | 5.29 |
| Norway | 0.97 | 6.19 | 4.24 | 1.65 |
| Portugal | 2.01 | 3.98 | 9.85 | 4.77 |
| Spain | 9.23 | 2.28 | 40.70 | 3.10 |
| Sweden | 1.78 | 6.74 | 7.91 | 1.77 |
| Switzerland | 1.60 | 12.50 | 5.68 | 5.99 |
| UK | 11.78 | 9.76 | 57.66 | 1.63 |
| 18 nations | 76.30 | 9.16 | 327.90 | 3.95 |

*Source*: OECD, 1988b, table C.2. Based on OECD Demographic Data File

would have taken over the lead with an increase of 72.5 per cent, double that of even Greece, Romania or Poland (table 8.4). In 1970 this region was clearly the youngest of the four European macro-regions, but by 2010 it may be slightly older than the European population in general. Greece and Italy in particular are moving quickly to join the countries with the highest proportions of their populations in the elderly age-groups. Since 1980 most members of this group have had high rates of very old population growth and these will continue during the first quarter of the next century, particularly in Italy.

Throughout Europe there will be a resurgence in the growth of the elderly population from 2010 until the early 2030s, as the large birth cohorts of 1945–70 reach old age. Among the 18 members of the OECD in Europe, the population aged 65 and over is expected to fall by 7 million in the 2040s (table 8.5). The only country which is unlikely to experience a decrease is Eire.

### Mortality in Late Age and Forecasts of the Elderly Population

The high rates of increase of the elderly population of Europe during the first half of this century were due to the relatively high birth rates and declining infant mortality in the second half of the nineteenth century. The slow down in the growth of the elderly population in the second half of this century has resulted from the lower birth rates in the early part of this century. The continued increase in the elderly population and the prospects for the first decades of the next century will soon be influenced more forcefully by trends in mortality in old age. These trends currently display considerable and rarely appreciated variations among European countries. Mortality in old age is exceptionally high in some East European countries: indeed the 1985 death rate of around 3 per cent for males aged 60–4 years in Hungary, Czechoslovakia and Poland was double that of Switzerland, Iceland and Sweden (table 8.6). It is also high in the British Isles and Finland: rates of 1.5 and 1.3 per cent among women aged 60–4 in 1985 for Scotland and Eire were more than double those for Switzerland.

During most of this century increases in average life expectancy at birth in European countries have been influenced more by the decline of infant and childhood mortality than by mortality in adult life. The likelihood of dying in the first ten years of life in England and Wales in 1987–9 was 0.013 for men and 0.010 for women, respectively only 6.5 and 6.1 per cent of the probability 75 years earlier, whereas the probability of dying during the seventh decade of life was 0.229 for men and 0.135 for women, respectively 64 and 47 per cent of the likelihood in 1910–12 (Benjamin, 1988, table 5). Indeed from 1901 to 1940, life expectancy at age 75 decreased from 6.4 to 6.0 years for men and remained at 7.2 years for women.

But by the late 1980s, in northern and western Europe around 95 per cent of deaths occurred to individuals aged 50 and older, and very recently mortality reductions in late-middle and early-old age have made substantial contributions to improving survival. Between 1891 and 1951 in the UK, male life expectancy at birth increased by 58 per cent to 66.2 years whereas at age 60 it improved by less than 25 per cent to 14.8 years (table 8.7). However, from 1951 to 1985 the position reversed with life expectancy at birth for men improving by 8.3 per cent to 71.7 years and at age 60 by 12.7 per cent to 16.6 years. There has been a comparable switch-over for females, but the gains have been much greater and so the female to male advantage at age 60 has

**Table 8.6** The level in 1985 and rate of change during 1970–85 of mortality or 60–4-year-old persons among European nations

| | *Death rate among 60–4 year olds*[a] | | | | *Annual change, 1970–85* | | | |
|---|---|---|---|---|---|---|---|---|
| | *Males* | | *Females* | | *Males* | | *Females* | |
| *Rank* | *Country* | *Rate* | *Country* | *Rate* | *Country* | *Rate of change* | *Country* | *Rate of change* |
| 1 | Hungary | 324 | Hungary | 148 | Hungary | −1.44 | Hungary | −0.31 |
| 2 | Czechoslovakia | 315 | Scotland | 147 | Bulgaria | −1.22 | Poland | −0.06 |
| 3 | Poland | 291 | Bulgaria | 139 | Poland | −0.86 | Bulgaria | 0.04 |
| 4 | Scotland | 263 | Czechoslovakia | 135 | Czechoslovakia | −0.40 | Denmark | 0.20 |
| 5 | Bulgaria | 254 | Poland | 129 | Denmark | −0.23 | Scotland | 0.29 |
| 6 | Finland | 250 | Eire | 129 | Romania | −0.22 | England and Wales | 0.29 |
| 7 | N. Ireland | 247 | Romania | 129 | Greece | −0.07 | Romania | 0.29 |
| 8 | Romania | 240 | N. Ireland | 126 | Yugoslavia | 0.29 | Norway | 0.36 |
| 9 | Eire | 238 | Yugoslavia | 126 | Eire | 0.30 | Czechoslovakia | 0.50 |
| 10 | Yugoslavia | 238 | GDR | 125 | Norway | 0.41 | GDR | 0.62 |
| 11 | GDR | 231 | England and Wales | 121 | Sweden | 0.56 | Eire | 0.86 |
| 12 | England and Wales | 218 | Denmark | 118 | GDR | 0.64 | N. Ireland | 1.22 |
| 13 | Luxembourg | 217 | Luxembourg | 101 | Italy | 0.73 | Yugoslavia | 1.22 |
| 14 | Denmark | 214 | Austria | 98 | N. Ireland | 1.12 | Greece | 1.32 |
| 15 | Austria | 209 | Belgium | 94 | Scotland | 1.14 | Netherlands | 1.59 |
| 16 | Belgium | 209 | FRG | 93 | England and Wales | 1.24 | Sweden | 1.43 |
| 17 | Italy | 204 | Portugal | 92 | Spain | 1.29 | Luxembourg | 1.69 |
| 18 | FRG | 203 | Italy | 88 | Portugal | 1.31 | Italy | 1.86 |
| 19 | France | 198 | Finland | 87 | Netherlands | 1.32 | Portugal | 1.88 |
| 20 | Portugal | 193 | Norway | 85 | Luxembourg | 1.47 | Austria | 2.00 |
| 21 | Netherlands | 186 | Greece | 85 | France | 1.49 | FRG | 2.11 |
| 22 | Norway | 186 | Netherlands | 81 | Finland | 1.61 | France | 2.21 |
| 23 | Greece | 175 | Sweden | 78 | Switzerland | 1.73 | Switzerland | 2.28 |
| 24 | Spain | 170 | France | 75 | FRG | 1.79 | Iceland | 2.33 |
| 25 | Sweden | 163 | Spain | 75 | Austria | 2.00 | Belgium | 2.33 |
| 26 | Iceland | 160 | Iceland | 73 | Belgium | 2.00 | Spain | 2.69 |
| 27 | Switzerland | 153 | Switzerland | 67 | Iceland | 2.19 | Finland | 2.71 |

[a] Deaths per 10,000 persons per year.

**Table 8.7** Life expectancy at birth and in old age (years): UK, 1891, 1951 and 1985

| | *At birth* | | | *At 60 years* | | | *At 75 years* | | |
|---|---|---|---|---|---|---|---|---|---|
| *Period* | *Males* | *Females* | *Ratio* | *Males* | *Females* | *Ratio* | *Males* | *Females* | *Ratio* |
| *Life expectancy* | | | | | | | | | |
| 1891 | 41.9 | 45.7 | 1.09 | 11.9 | 13.0 | 1.09 | 5.6 | 6.1 | 1.09 |
| 1951 | 66.2 | 71.2 | 1.08 | 14.8 | 17.9 | 1.21 | 6.7 | 8.0 | 1.19 |
| 1985 | 71.7 | 77.4 | 1.08 | 16.6 | 21.0 | 1.37 | 7.9 | 10.5 | 1.33 |
| *Gain in life expectancy* | | | | | | | | | |
| 1891–1951 (% gain) | 58.0 | 55.8 | | 24.4 | 37.7 | | 19.6 | 31.1 | |
| 1951–85 (% gain) | 8.3 | 8.7 | | 12.7 | 17.3 | | 17.9 | 31.3 | |
| 1891–1951 (% p.a.)[a] | 0.77 | 0.74 | | 0.36 | 0.53 | | 0.30 | 0.45 | |
| 1951–85 (% p.a.)[a] | 0.24 | 0.25 | | 0.34 | 0.47 | | 0.49 | 0.80 | |

[a] Average annual compound change as a percentage.
*Source*: Preston et al., 1972

increased from 9 per cent in 1891 to 37 per cent in 1985. The annual rates of improvement in the life expectancies at different ages neatly summarize the growing importance of improving survival in old age. During 1891–1951, life expectancy at birth among women increased by an average of 0.74 per cent per year, but afterwards the rate decreased by two-thirds to 0.25 per cent. But the expectancy of remaining life at 75 years increased by 0.45 per cent per year in the earlier period and subsequently almost doubled to 0.80 per cent (table 8.7).

Recent improvements in mortality in later life have led to repeated under-estimation of the increase in the population of very old people in the USA and the UK (Alderson and Ashwood, 1985; Benjamin, 1988; Olshansky, 1988; Warnes, 1989b). There has been a further acceleration of mortality improvement in the UK during the 1980s (Warnes, 1992b). The prospect of a prolonged period of improvement among children would be welcomed without qualification, but for old people its joint effect with very low fertility generates concern. Demographers at the OECD, for example, have produced population projections with the optimistic mortality assumption of an increase in life expectancy at age 60 of ten years for men and women between 2000 and 2030 and stabilization thereafter (OECD, 1985, p. 17). The revised assumption produces age structures as yet unseen, with those aged 65 and over reaching one-fifth of the total population in 2020 and about one-third by mid-century in France, 36 per cent in Germany, 31 per cent in Italy and 29 per cent in the UK.

Recent actual trends in mortality in old age show substantial contrasts among European countries, even within the Community. Too much should not be made of individual slight differences, for some artificial perturbations can be expected to result from age misreporting on death certificates and other flaws in registration. The secular decline of mortality in later ages may be intrinsically irregular (Brody, 1985).

It is far from clear what causes this dynamism and variability. Specific analyses have established links between some patterns of behaviour, notably tobacco smoking, and certain differential aspects of the trends, such as that between men and women (Alderson and Ashwood, 1985). But a host of other life-style, occupational, nutritional, environmental and even foetal conditions may be implicated, as well as the advances in medical technologies and treatments in the prevention and delay of deaths from cardiovascular diseases and neoplasms. Examination of these trends has generated intense research and public policy interest in the USA and Japan, partly because of the uncertainty over the relationship between improving survival and the age distribution of disability and morbidity (Fries, 1980, 1989). The reduction of premature deaths has produced a more rectangular survival curve with progressively higher percentages of successive cohorts surviving to older ages. The current debate is concerned with the issue of whether this is resulting in a reduction in the variance in age at death, the compression of morbidity into a few years around the modal age at death and therefore a future increase or decrease in the per capita years of dependence and the effects on the need for health care. These matters have received attention from national and international agencies in Europe, notably the OECD and the World Health Organization, but a fundamental difficulty facing further research work is the inadequate age-specific health data for European populations.

Details of the changes in mortality in old age since 1970 are indicated by the change in death rates among those aged 60–4 for seven European countries (table 8.8). This shows the large and widening differentials between men and women among the countries. In 1970, the coefficients of variation of life expectancy for men were 2 per cent at birth, 4.5 per cent at age 60 and around 5.3 per cent at age 75. The equivalent variations for women were 1.5, 3.1 and 4.9 per cent. By the mid-1980s the coefficients had widened. For men they were 2.2, 6.8 and 8.1 per cent, and for women 2.3, 6.4 and 8.4 per cent. The contrasts are heightened by the widening differentials between the Dutch and Polish rates, which for life expectancy among 75-year-old women widened over the period from a 14 per cent to a 28 per cent disadvantage for the Poles. Comparisons among social groups and

**Table 8.8** Life expectancy (years) at ages, 0, 10, 60 and 75 years, 1970–85

| | | *At birth* | | *At 10 years* | | *At 60 years* | | *At 75 years* | |
|---|---|---|---|---|---|---|---|---|---|
| | | *Males* | *Females* | *Males* | *Females* | *Males* | *Females* | *Males* | *Females* |
| Switzerland | 1970 | 69.5 | 75.9 | 62.2 | 68.1 | 17.1 | 20.8 | 8.2 | 9.7 |
| | 1986 | 73.8 | 80.6 | 65.0 | 71.8 | 19.1 | 24.1 | 9.4 | 12.1 |
| | % change[a] | 6.2 | 6.3 | 4.5 | 5.4 | 11.9 | 15.6 | 15.4 | 24.8 |
| Spain | 1970 | 69.4 | 74.9 | 62.1 | 67.2 | 17.1 | 20.2 | 8.1 | 9.4 |
| | 1983 | 72.9 | 79.2 | 64.1 | 70.2 | 18.5 | 22.4 | 8.9 | 10.7 |
| | % change | 5.1 | 5.7 | 3.2 | 4.5 | 8.1 | 10.9 | 9.4 | 14.3 |
| France | 1970 | 68.6 | 76.1 | 60.2 | 67.5 | 16.2 | 20.9 | 7.8 | 9.9 |
| | 1985 | 71.3 | 79.6 | 62.2 | 70.4 | 17.9 | 23.2 | 8.6 | 11.3 |
| | % change | 3.9 | 4.5 | 3.3 | 4.3 | 10.3 | 10.7 | 10.0 | 13.4 |
| Poland | 1970 | 66.4 | 73.0 | 59.6 | 65.7 | 15.5 | 19.0 | 7.3 | 8.5 |
| | 1985 | 66.5 | 74.8 | 58.2 | 66.2 | 15.2 | 19.5 | 7.1 | 9.0 |
| | % change | 0.1 | 2.4 | −2.4 | 0.8 | −2.5 | 2.7 | −3.3 | 5.2 |
| Denmark | 1970 | 70.9 | 76.1 | 62.5 | 67.2 | 17.2 | 20.6 | 8.5 | 10.0 |
| | 1985 | 71.5 | 77.6 | 62.4 | 68.3 | 17.2 | 21.7 | 8.4 | 11.0 |
| | % change | 0.9 | 2.0 | −0.2 | 1.7 | −0.0 | 5.6 | −1.4 | 10.3 |
| Netherlands | 1970 | 70.8 | 76.6 | 62.3 | 67.7 | 16.9 | 20.6 | 8.3 | 9.7 |
| | 1985 | 73.1 | 79.8 | 63.9 | 70.6 | 17.6 | 23.0 | 8.4 | 11.5 |
| | % change | 3.1 | 4.3 | 2.6 | 4.2 | 4.3 | 11.6 | 1.5 | 18.8 |
| UK | 1970 | 68.7 | 75.0 | 60.5 | 66.5 | 15.3 | 19.8 | 7.4 | 9.5 |
| | 1985 | 71.7 | 77.4 | 62.6 | 68.3 | 16.6 | 21.0 | 7.9 | 10.5 |
| | % change | 4.3 | 3.2 | 3.5 | 2.6 | 8.6 | 5.9 | 6.7 | 10.2 |
| Coefficient | 1970 | 2.04 | 1.49 | 1.81 | 1.11 | 4.52 | 3.08 | 5.34 | 4.85 |
| of variation[b] | 1980 | 3.12 | 2.33 | 3.26 | 2.52 | 6.82 | 6.44 | 8.10 | 8.35 |

[a] Increase calculated from expectancies to three decimal places.
[b] The coefficient of variation is the standard deviation of the seven national values divided by the mean.

with other countries indicate considerable scope for further improvement. Age-specific death rates at 50–74 years are at least 30 per cent lower in Japan than in the UK (Warnes, 1992a).

### Secondary Demographic Characteristics

In most of Europe, the later phases of demographic ageing have been accompanied by strong economic growth and many social changes. Increasing survival has been accompanied by intricate changes in marital status in old age. Today's young elderly population, born after 1918, have generally experienced a higher marriage rate than the preceding cohorts. The proportion of young elderly people who are married has been rising, while the proportion widowed has been falling. At older ages, however, the increasing differential between men and women in terms of survival has been increasing the proportion of women who are divorced. The increase in divorce rates among those in mid-life during the 1960s will affect living arrangements during the next few decades.

Economic growth can be associated with two distinct parallel trends: the decreasing participation in paid work at older ages and rising household headship rates. The former has been reinforced by structural changes, particularly the declining importance of employment in agriculture, and it has been facilitated by the rising real value and 'reach' of state and occupational pensions. In France, the economic activity rate among men aged 60–4 has fallen from 70 per cent in 1954 to 63 per cent in 1970, 48 per cent in 1980 and 30 per cent in 1985. In West Germany, the equivalent rates have fallen from 75 per cent in 1970 to 43 per cent in 1980 and 33 per cent in 1985 (Walker et al., 1991, table 6). This trend is as much imposed by employers as it is a reflection of voluntary behaviour, but it is of course converting large numbers of senior employees from taxpayers to social security beneficiaries. Innumerable analyses of these specific trends have documented the implications for national accounts and in a few countries they have led to the curtailment of entitlements and revisions to the age of eligibility for benefits and pensions (Rix and Fisher, 1982; Wilson and Wilson, 1991). Analyses of the distribution of income in old age have also been undertaken. They tend to show the increased prevalence of poverty in old age, not least among the countries of the EC.

The tendency for elderly people to live independently in small households is a particular characteristic of societies in north-western Europe, while in the south and east multi-generational households are still relatively common. A recent comprehensive survey has estimated that median household size in the member states of the Council of Europe (1990a) fell from 3.4 in 1960 to 3.1 in 1970 and 2.7 in 1980. Sweden, Switzerland, Denmark and Germany had household sizes of 2.5 or less in 1980 and Cyprus, Greece, Eire, Italy, Malta, Portugal, Spain and Turkey all had mean sizes of more than three persons. These variations reflect differences in fertility and family size, but they probably also indicate the extent to which elderly people are likely to live on their own. Even in Hungary, the percentage of men aged 60 and over living with others apart from their spouse or partner has fallen from 14 in 1960 to seven in 1984, and among women of the same age the decline has been from 40 to 30 (Vukovich, 1991, table 12).

## The Problems of Demographic Ageing in Europe

There are profound health service and welfare problems among the disadvantaged and sick minority of Europe's elderly population. Living standards remain low in many parts of eastern and southern Europe, particularly in rural areas and among ethnic minorities in the cities. Locally in eastern Europe, difficulties have been much compounded in the last few months by the political stresses and changes following the collapse of state communism. In the countries of the north west with welfare state systems, a long-term consensus that the most disadvantaged should be supported by the progressive elaboration of universal income support schemes and state-managed residential and domiciliary services has been challenged with increasing vigour by the new-right parties. There is increasing advocacy that further improvements in general well-being should be more the responsibility of market mechanisms, individual responsibility and choices, and the private sector.

This debate represents the first major reappraisal of European welfare state principles since the Second World War. But its relation to demographic ageing is tenuous, even if simpler contributions to the arguments cynically use the prospective growth of the elderly population as an indicator of impending economic ruin. If in the next decade European fertility rose well above replacement level and our population began to rejuvenate, the fundamental issues relating to the financing and management of income support and welfare services would not change. These issues are problems of politics, ideology, social administration and sociology before they are problems of demography.

The key demographic problems associated with ageing are how fast it is occurring and how large the elderly population's share will become. The principal control is the level of fertility, but another influence is the level of mortality in later life. If fertility does not recover and mortality continues to improve, then by the middle of the next century the elderly population will make up as much as one-third of the total.

Of course, rising fertility or constant or even rising mortality could subvert these projections, but if they are proved right the outcome should be celebrated more than feared. It will only come about if morbidity and mortality conditions improve, which by definition implies that people in their 60s and 70s will be living healthier lives than they are today. The elderly population of the future will also be more affluent, better educated and better housed. There is no reason to believe that they will not make diverse and full contributions to their local and national communities. The elderly population may grow substantially in both absolute and relative terms and their health expectations will almost certainly continue to expand, but in neither fiscal nor sociological terms is it proven or even likely that they will be an increasing burden on European society.

# 9

# Family Structures

Ray Hall

The last 20 years have been a period of great change for families and households in Europe and there is now widespread debate on the future of the family: will it survive in its present form? This chapter outlines some of the key changes that have been taking place within both households in general and families in particular. After looking briefly at sources of data, the discussion will include the following: first, a look at changes in household size during the last three decades and how these trends have affected the structure of the family; second, a look at the related processes that are involved in the changes, especially changes in marriage and divorce, the growth of cohabitation, the decline of fertility and the changing role of women, specifically the increase of waged work among women; and, third, a brief review of policy issues.

Household and family formation is a particular point of reference at which demographic phenomena and economic, social, psychological and cultural processes interact. An insight into the changing patterns and structures of households and families is of great importance to policy-makers and analysts in many fields since the household is one of society's basic consumption units. Many key decisions, including consumption decisions, are taken at the household level rather than by individuals.

The term family is used to mean the nuclear family, united by blood, marriage or adoption and sharing a common residence (United Nations, 1973). Until relatively recently the terms family and household were sometimes used synonymously. For example, in the 1981 Italian census the word *famiglie* was used to denote households. However, the forms of households are becoming increasingly varied and the family can no longer be used as an all-encompassing term for households. Three broad categories of households may be identified: single family households, multiple family households and non-family households.

### Sources of Data

The population census is the prime source of information about the household, although increasingly sample surveys are providing valuable information concerning changes in the family and household. More refined sources, such as the British Longitudinal Survey, offer considerable potential for detailed analysis of the processes and changes of households (Murphy et al., 1988).

Problems of terminology abound, not least in the commonly used yet now anachronistic term 'head of household' which assumes an economic unit with one person making the major contribution. In Belgium in 1971 the 'head' was defined as the person who exercised the greatest authority (Hall, 1986). This form of definition no longer reflects the reality of normal family circumstances because it fails to recognize that women, whether in paid work or not, make significant and equal contributions to the household. Such a concept is also irrelevant in non-family households. Most censuses now use the notion of a 'reference person', although even this term is subject to mutation.

### The Changing Size of the Household

A consideration of changing household size provides the starting point for an analysis of the processes involved. Despite differences of definition and problems of data comparability, trends in household size are remarkably consistent.

The mean household size (MHS) provides the most straightforward single-number index of household structure (table 9.1). Around 1960 the average MHS for the 12 European Community (EC) countries was 3.36. Only two countries, the Federal Republic of Germany and Denmark, had an MHS under 3, while Eire's was the highest at 3.96. By 1970 the average MHS had declined to 3.23; it was less than 3 in four countries, but it was still highest in Eire with 3.93. Decline of the MHS was much more rapid in the 1970s and around 1980 the average MHS was 2.93 with seven countries below 3 and Eire at 3.66. Over the three decades the range of MHS among the 12 EC countries widened from 1.08 in 1960 to 1.26 in 1980.

Between 1961 and 1971 the fastest decline in MHS occurred in Greece, the Netherlands, the UK and Italy, all with a fall of 0.3

**Table 9.1** Mean household size around 1960, 1970 and 1980 and absolute change during 1960–70 and 1970–80

| *Country* | *1960* | *1970* | *1980* | *Absolute change, 1960–70* | *Absolute change, 1970–80* |
|---|---|---|---|---|---|
| Belgium | 3.01 | 2.95 | 2.67 | 0.06 | 0.28 |
| Denmark | 2.90 | 2.70 | 2.40 | 0.20 | 0.30 |
| Eire | 3.96 | 3.93 | 3.66 | 0.03 | 0.27 |
| France | 3.10 | 2.88[a] | 2.70[b] | 0.22 | 0.18 |
| Germany (FRG) | 2.88 | 2.74 | 2.43 | 0.14 | 0.31 |
| Greece | 3.78 | 3.39 | 3.12 | 0.39 | 0.27 |
| Italy | 3.63 | 3.35 | 3.01 | 0.28 | 0.34 |
| Luxembourg | 3.21 | 3.07 | 2.79 | 0.14 | 0.28 |
| Netherlands | 3.58 | 3.20 | 2.76 | 0.38 | 0.44 |
| Portugal | 3.72 | 3.67 | 3.35 | 0.05 | 0.32 |
| Spain | – | 3.81 | 3.53 | – | 0.28 |
| UK | 3.21 | 2.91 | 2.72 | 0.30 | 0.19 |

[a] 1975.
[b] 1982.
*Source*: Council of Europe, 1990a, appendix table 3A

or more. Between 1971 and 1981 the UK experienced the least change, only 0.19, while the Netherlands – closely followed by Italy, Portugal and the Federal Republic of Germany – continued to experience a rapid rate of decline. It is clear that in spite of the continuing variations in MHS among the 12 countries the trends are all in the same direction, and by the early 1990s it is likely that all countries in the EC will have an MHS below 3.

The decline in MHS is a reflection of the changing size distribution of households in the 12 countries. Two trends are particularly notable: fewer large households and the rapid increase of single-person households (table 9.2). By 1981 households of five or more persons made up a shrinking minority. Only in Eire, with 32 per cent, Spain with 26 per cent and Portugal with 21 per cent did they comprise a significant minority. Elsewhere, they accounted for around 10 per cent of households, with only 7 per cent in Denmark. On the other hand, one-person households increased rapidly in importance after 1961 and by 1981 they were approaching or had exceeded 20 per cent of all households in a majority of EC countries and nearly a third of households in Denmark and the Federal Republic of Germany. Only in Spain and Portugal did proportions remain relatively small: 10 and 13 per cent respectively.

**Table 9.2** Size distribution of households around 1970 and 1980

| | *Total number of households* | *Distribution of households by size (%)* | | | | |
|---|---|---|---|---|---|---|
| | | *1* | *2* | *3* | *4* | *5* |
| Belgium | | | | | | |
| 1970 | 3,234,228 | 18.8 | 30.2 | 20.1 | 14.8 | 16.1 |
| 1981 | 3,608,178 | 23.2 | 29.7 | 20.0 | 15.7 | 11.4 |
| Denmark | | | | | | |
| 1970 | 1,790,815 | 20.8 | 30.0 | 19.2 | 17.3 | 12.7 |
| 1981 | 2,028,516 | 29.0 | 31.5 | 15.9 | 16.2 | 7.4 |
| Eire | | | | | | |
| 1971 | 730,543 | 14.2 | 20.6 | 16.0 | 14.1 | 35.2 |
| 1981 | 910,700 | 17.1 | 20.2 | 15.0 | 15.4 | 32.3 |
| France | | | | | | |
| 1968 | 15,762,508 | 20.3 | 26.9 | 18.6 | 15.0 | 19.1 |
| 1982 | 19,588,924 | 24.6 | 28.5 | 18.8 | 16.2 | 11.9 |
| Germany (FRG) | | | | | | |
| 1970 | 21,991,000 | 25.1 | 27.1 | 19.6 | 15.2 | 12.9 |
| 1980 | 25,336,000 | 31.3 | 28.7 | 17.7 | 14.4 | 8.0 |
| Greece | | | | | | |
| 1971 | 2,491,916 | 11.3 | 21.4 | 21.2 | 23.9 | 22.1 |
| 1981 | 2,974,450 | 14.6 | 24.7 | 20.2 | 24.0 | 16.5 |
| Italy | | | | | | |
| 1971 | 15,981,177 | 12.9 | 22.0 | 22.4 | 21.2 | 21.5 |
| 1981 | 18,632,337 | 17.8 | 23.6 | 22.1 | 21.5 | 14.9 |
| Luxembourg | | | | | | |
| 1970 | 108,498 | 15.7 | 27.1 | 22.0 | 18.1 | 17.1 |
| 1981 | 128,281 | 20.7 | 28.5 | 21.2 | 17.5 | 12.1 |
| Netherlands | | | | | | |
| 1971 | 3,990,000 | 17.1 | 25.3 | 18.1 | 19.1 | 20.4 |
| 1981 | 5,111,000 | 22.1 | 29.9 | 15.6 | 20.7 | 11.7 |
| Portugal | | | | | | |
| 1971 | 2,345,225 | 10.0 | 21.9 | 22.3 | 18.5 | 27.3 |
| 1981 | 2,907,402 | 12.9 | 23.9 | 22.9 | 20.0 | 20.7 |
| Spain | | | | | | |
| 1970 | 8,853,660 | 7.5 | 18.0 | 19.2 | 21.8 | 33.5 |
| 1981 | 10,586,441 | 10.2 | 21.4 | 19.8 | 22.2 | 26.4 |
| UK | | | | | | |
| 1971 | 18,541,964 | 18.1 | 31.4 | 19.1 | 16.9 | 14.5 |
| 1981 | 19,948,776 | 21.7 | 31.8 | 17.1 | 18.1 | 11.4 |

*Source*: Council of Europe, 1990a, appendix table 3A

The reduction in the number of large households and the increase in one-person households reflect dramatic changes in the structure of the family. Multi-generational households were declining rapidly, the number of children per household and the number of households with children were in decline and increasing numbers of elderly persons were living alone. At the same time, young adults were leaving the parental home and living alone for at least some time before joining with others in a variety of household arrangements. Two- and three-person households experienced rather less change, although there was a tendency for the former to increase while the latter declined.

Contrasts between northern and southern members of the Community have persisted but they are by no means uniform. By 1981 Italy had a relatively small proportion of large households and an increasing number of one-person households, although those with two or three persons were still the most common. In Greece, Spain and Portugal the process of change was less advanced with fewer single-person and more large households, especially in Spain. Eire occupied an anomalous position with relatively large numbers of one-person households (17 per cent) while at the same time having nearly a third of households with five or more persons.

Certainly, what is often taken to be the stereotypical household comprising husband, wife and two children can no longer be regarded as the norm in the Community today since in no country do four-person households make up more than 24 per cent of the total. Even so the majority of people in the EC still live in households headed by or made up of a married couple. In the case of the UK, a country where the decline of large households is already at an advanced stage, 79 per cent of people do so (Henwood et al., 1987).

Trends in household size point to a process of individualization, of household fission, leading to large numbers living alone or with just one other person, and to the increasing rarity of multi-generational households. This process also implies a reduction in family households, that is, households with dependent children. One consequence is that the majority of adults now live in households without dependent children and increasingly they live alone. Only in Eire, Spain and Portugal do more than half of all adults live in households containing children. In the UK, Denmark, France, Belgium and the Netherlands more than 10 per cent of adults live alone, while under 7 per cent live alone in Greece, Eire, Portugal and Spain.

These simple figures, however, obscure from view other structural changes that have been taking place. For example, the increase in marital breakdown has made more common complex patterns of one-parent households and remarried couples with stepchildren; and this has occurred alongside the growth of consensual unions and the decline in the numbers of both relatives and non-relatives living with families.

## Family and Household Changes: Social and Demographic Causes and Effects

Demographic changes and trends in nuptiality show a remarkable degree of convergence throughout the EC. The period since 1945 has been characterized by a rise in life expectancy, economic growth and an increase in the cost of having children and, from the mid-1960s, fertility has declined coinciding with the first generation born after the war to reach reproductive age. The behaviour of this generation was different, a result perhaps of their very different upbringing, a revolution in education and the changing role of women. Everywhere in the Community, despite substantial differences that exist among countries in the sociocultural sphere and with respect to legislation – especially that relating to divorce – third births have been declining, marriages are taking place later, births outside marriage and cohabitation are increasing and the divorce rate is increasing.

> All of these changes are associated with the arrival to adulthood of the post-Second World War generation. There seems to be a fundamental change in attitudes to life which is not caused by political or economic factors peculiar to a region or social background. There would seem to be a major transformation affecting all advanced countries, irrespective of their social and regional structure, and the movement is universal, causes are therefore deep-seated, common to all the Western countries and capable of changing mentalities.
> (Burnel, 1986, p. 9)

The major changes that have been affecting the size and structure of households and families include first, changing behaviour towards marriage; second, the growth of cohabitation outside marriage; third, an increase in the frequency of divorce; fourth, the decline of fertility within marriage; and, fifth, an increase in fertility outside marriage. The last two mentioned are dealt with at greater length in other chapters, but the first three will be considered here.

### *Marriage*

Marriage is the traditional institution by which families are established, yet the last 20 years have seen substantial changes in the rate at which people are marrying. It is no longer the necessary first step in family formation.

From a peak in the mid-1960s, the number of marriages has declined in every Community country (see figure 9.1). Although the marriage rate per thousand total population is a crude measure of the extent of marriage, it does give an indication of broad trends in the rate at which people are marrying in different countries. The crude marriage rate for the EC as a whole has shown an overall decline since 1960. Between 1960 and 1973 rates were relatively stable and varied between 7.5 and 7.9 per thousand. By 1973 it was 7.5 and declined steadily thereafter reaching 6.3 by 1980 and a low point of 5.7 between 1984 and 1986. In 1987 and 1988 it rose to 5.9.

There were variations in trend between individual countries (table 9.3). In 1960 there was considerable variation in crude marriage rates which ranged from 5.5 in Eire, reflecting the continuation of traditional marriage patterns, to 9.4 in West Germany. The other ten countries varied between 7.0 in Greece and France and 7.8 in Denmark, the Netherlands and Portugal. By 1970 the Netherlands had the highest marriage rate at 9.5 and Luxembourg the lowest with 6.3. Eire was the next to lowest with 7.3, but this was now far closer to the average for the rest of Europe. By 1980 every country of the EC had experienced a decline from the 1970 figure and the range was much reduced, from 5.2 in Denmark to 7.4 in the UK. The decline continued from 1980 to the mid-1980s. Since then most countries have experienced a slight increase in marriage rates. Greece recorded the lowest marriage rate in 1988 with 4.8. This may reflect the Greek custom of avoiding marriage in a leap year because of the belief that such a marriage will suffer

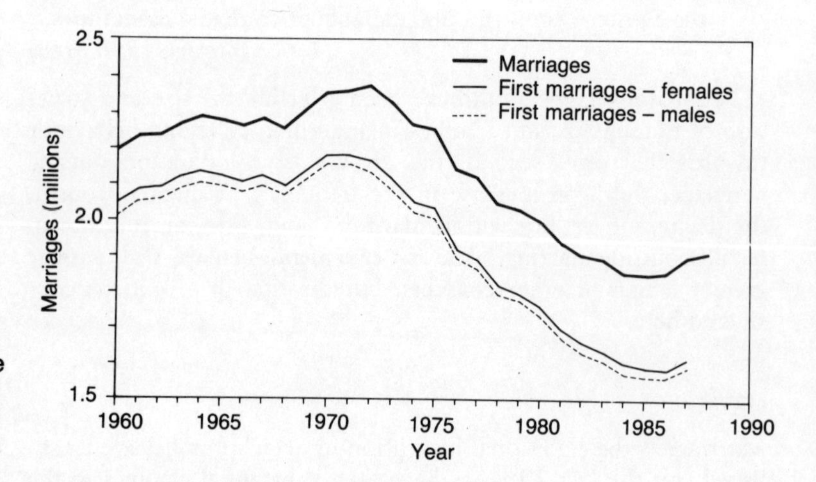


**Figure 9.1** Trend in the number of marriages in the European Community.

**Table 9.3** Gross marriage rate in European Community countries, 1960–88

| | *1960* | *1970* | *1980* | *1985* | *1988* |
|---|---|---|---|---|---|
| Belgium | 7.2 | 7.6 | 6.7 | 5.8 | 6.0 |
| Denmark | 7.8 | 7.4 | 5.2 | 5.7 | 6.3 |
| Eire | 5.5 | 7.0 | 6.4 | 5.3 | 5.1 |
| France | 7.0 | 7.8 | 6.2 | 4.9 | 4.9 |
| Germany (FRG) | 9.4 | 7.3 | 5.9 | 6.0 | 6.5 |
| Greece | 7.0 | 7.7 | 6.5 | 6.4 | 4.8 |
| Italy | 7.7 | 7.3 | 5.7 | 5.2 | 5.5 |
| Luxembourg | 7.1 | 6.3 | 5.9 | 5.3 | 5.5 |
| Netherlands | 7.8 | 9.5 | 6.4 | 5.7 | 6.0 |
| Portugal | 7.8 | 9.0 | 7.4 | 6.7 | 6.8 |
| Spain | 7.7 | 7.3 | 5.9 | 5.2 | 5.5 |
| UK | 7.5 | 8.5 | 7.4 | 6.9 | 6.9 |

*Source*: EUROSTAT, 1990, table F2

bad luck (Festy, 1983). By 1988 the marriage rate was at its highest level in the UK and Portugal (6.9 and 6.8 respectively) where rates had not fallen to the same extent as in other parts of the Community. More remarkable was the steady increase in Denmark's marriage rate from a low of 5.2 in 1980 to 6.3 in 1988. In the Federal Republic of Germany the rate had risen to 6.5 (EUROSTAT, 1990).

The overall picture reflects a decline in first marriage rates from the 1970s onwards, with a slight recovery in the last two or three years. Even so, there seems little likelihood that marriage rates will increase to their 1960s levels. However, recent changes in the Swedish marriage rate should be noted. There the rate rose spectacularly in 1989 when there was a total of 108,000 marriages compared with only 44,000 in 1988. This rise could well be a precursor of changes elsewhere in Europe, although the immediate cause in Sweden's case was a law excluding childless cohabiting women from entitlement to their partner's pension (*Population et Société*, 1990).

At the same time that marriage rates were declining, the mean age at marriage was rising. Among the 12 members of the EC the mean age at marriage was 28.5 for males and 25.3 for females in 1960. By 1970 these had declined to 27.4 for males and 24.4 for females; they then rose steadily to reach 29.5 and 26.6 respectively in 1987. Mean age at first marriage also declined from 1960 (26.9 and 24.1) to 1975 (25.6 and 23.0); thereafter it increased to 27.1 and 24.6 in 1987.

The difference between these two means results from the influence of second and subsequent marriages and since these have increased in number during the period being discussed, so has the difference between the two measures. Since the mean age at first marriage avoids the distorting influence of remarriage, its use is probably more appropriate for international comparisons.

The timing of changes in age at first marriage was not uniform throughout the EC. The four countries of southern Europe, together with Eire, lagged behind the other seven so that there the decline in mean age at first marriage continued to 1980 rather than 1975. Although it has since increased, in almost all these countries the mean age at first marriage is still slightly lower than it was in 1960.

In 1960 the highest mean age at first marriage for males and females was in Eire, 30.8 and 27.1 respectively, reflecting the persistence of a traditional marriage pattern reminiscent of the nineteenth century. The youngest mean age at first marriage for males was 25.7, in the UK, and for females was 22.9, in Denmark. In the following five years age at first marriage declined, although the relative rankings were preserved. By 1970 the oldest mean age at first marriage for males was 27.9, in Greece, and for females was 24.8, in Eire, while the youngest age at first marriage for men had fallen slightly to 24.4, in Belgium and France, and had risen slightly for females to 22.4, in Belgium, France and the UK. By 1980 Denmark had the highest mean age at first marriage for both males and females, 27.5 and 24.8, and the youngest for men was 24.7, in Belgium, and for females 22.3, in Belgium and Greece. The pattern hardly changed in the next five years, although mean age at first marriage continued to increase. In 1988 Denmark still had the oldest age at first marriage, now 29.6 for males and 27.1 for females, while Portugal had the youngest age for males, 25.9, and Greece had the youngest age for females, 23.3.

Recent trends in the mean age at first marriage demonstrate some divergence. Increase has been particularly notable in countries like Denmark and the Federal Republic of Germany, where the change has been substantial, but there have also been increases in France, the Netherlands and the UK. In certain other countries – Greece, Portugal and Spain, for example – there has been persistent, if undramatic, decline.

The older age at first marriage still found in some countries, particularly those of southern Europe and Eire, in the early 1960s was characteristic of a traditional marriage pattern which had disappeared elsewhere by or just after the Second World War. This low nuptiality pattern has now given way to a generally younger age at marriage; meanwhile in northern Europe the younger age at marriage typical of the 1960s has been replaced by an older mean age at first marriage. However, unlike the earlier period of delayed marriage this older age at marriage does not coincide with delay in the formation of marital-type relationships.

Younger ages at marriage for males and females are now found in Portugal, Spain, Belgium and the UK, where men married before reaching 27 and women before 25, on average. At the other extreme come Denmark and the Federal Republic of Germany. Typically, the age difference between spouses in the Community is 2–3 years, apart from Greece where the difference is as much as 4.4 years. Greek women marry at the youngest age of any women in the EC, while the men are among the last to marry. It could be argued that this larger differential suggests a more traditional marital relationship with younger women marrying older, more dominant men. It is also probably the last vestige of the once distinctive southern European marriage pattern in which nearly all women got married, in their late teens whilst still virgins, to men who were substantially older, normally in their late 20s.

The general outlines of the new European marriage pattern seem relatively clear: not only is marriage becoming more unpopular, but when it does take place it is being delayed. However, this does not mean that people are not forming relationships, rather that the position of marriage as a regulator of sexual relationships is weakening. Willekens (1988) has argued that the security which the institution of marriage used to offer to the weaker partner has been undermined by changes in the law relating to marriage, by the increasing economic independence of women and by state social security provision. At the same time, increased individualism in society has resulted in weaker social ties which, together with changes in the social norms governing cohabitation, combine to weaken the position of marriage as an organizing principle for everyday life. Increasingly, types of explicit or implicit arrangements other than the legal marriage contract govern relationships between cohabiting people.

Legal changes with respect to marriage have been taking place since the 1970s in some southern European countries where up until then traditional marriages were enshrined in law. In Italy, such changes may be dated from the introduction of legal divorce in 1970 and were further encouraged by the reform of family law in 1975. This introduced a completely egalitarian regime between the spouses, ending the husband's legal status as head of the household and abolishing a wife's adultery as a penal offence (Donati, 1990). In Spain, in the same year, that aspect of the law which stated that 'husbands should protect their wives and these should obey them' was abolished. A new equality was later enshrined in Article 32 of the Spanish constitution: 'men and women have the right to marry with full legal equality.' Divorce became possible in 1978, although it is still difficult and costly to obtain (Del Campo, 1990).

In Greece, until 1983 women were not considered autonomous and independent individuals within the family and the patriarchal family was protected by law. Up to that time the husband was regarded as both head of the family and the person responsible for deciding all matters relating to the family. The wife's duties were confined to the care of the household and there was even a dowry obligation. Changes came with a new law in 1983 which established that there should be no distinction between the sexes concerning marriage, separation and divorce and men and women should have equal rights and obligations within the family. Even so, attitudes are changing only slowly (Symeonidou, 1990). These legislative developments have brought the Community's southern members closer to the northern experience, but underlying cultural differences with respect to marriage and the role of the family may be more persistent.

### *Cohabitation*

Twenty years ago cohabitation was a marginal phenomenon, but the fall in marriage rates experienced during the 1970s was accompanied by a rapid increase in non-marital cohabitation, particularly associated with the generations born after 1950. Since cohabitation, unlike legal marriage, is essentially private, it is likely to go largely unrecorded with the result that data on cohabitation are generally inadequate and what little information there is will have been gleaned from a wide variety of sources. In broad terms, cohabitation is much more frequent in northern countries of the Community than in the southern countries or Eire. Sweden has the highest level in Europe with nearly all marriages being preceded by a period of cohabitation. Within the EC Denmark has the highest rate. In Denmark in 1981 there were 449,000 cohabiting couples out of a population of 5 million, and 134,000 of these unions had lasted more than five years. Some 35 per cent of the 20–4 age-group live together and 45 per cent of all births take place outside marriage. In most respects cohabiting couples are considered to be married couples.

In France, from 411,000 consensual unions in 1975 the number rose to 710,000 in 1981, 5.4 per cent of all couples. Today, more than half of all couples cohabit before marriage and in the 20–4 age-group more than 30 per cent of all couples are not married, a proportion that increases to half in Paris (Roussel, 1987). A more detailed picture of cohabitation in France has been obtained from a survey conducted in 1985 by members of the Institut National d'Etudes Démographique (Leridon and Villeneuve-Gokalp, 1988). In 1968, 20 per cent of all unions began outside marriage and by 1985 this had increased to 65 per cent. The survey also revealed the variety of forms cohabitation had taken. It had originally been assumed that cohabitation was a preliminary to legal marriage, but it is now realized that for some couples cohabitation is an alternative to marriage. Types of cohabitation vary from temporary unions through to stable unions with children. The survey also showed that cohabitation was much more likely among individuals who had experienced the divorce of their parents and who were brought up by one parent. Daughters of working mothers were much more likely to want to maintain their independence and not marry. The increased popularity of cohabitation has in its turn weakened the pressures on couples to marry, so promoting its social acceptability even further (Villeneuve-Gokalp, 1990).

Cohabitation rates in the Federal Republic of Germany have been increasing since the early 1970s. The percentage of women who had experience of pre-marital cohabitation rose from 20 for those married before 1974 to 35 among those married in 1977–8 (Hopflinger, 1985). Not more than 10 per cent of consensual unions have children and in only a third of these is the child the offspring of both currently cohabiting partners (Schwarz, 1988).

In Great Britain, questions about cohabitation among women aged 18–49 have been included in the General Household Surveys since 1979. These data show a steady increase. In 1979 only 3 per cent of women were cohabiting at the time of interview; by 1988 the figure had risen to 8 per cent. Since 1986 questions about cohabitation have been asked of men and women aged 16–59 so that a much fuller picture may now be obtained. Among women who were not married and aged 18–49 the percentage cohabiting increased from 11 in 1979 to 19 in 1987. Within this group divorced women were much more likely to cohabit than never-married women. The percentage of divorced women cohabiting was 20 in 1979 and had risen to 27 in 1987, while among never-married women the percentage rose from 8 to 17 over the same period. For men the differential was even greater: 13 per cent of bachelors were cohabiting in 1987 compared with 38 per cent of divorcees. There has also been a substantial increase in the proportions of married men and women who reported that they had cohabited before marriage. For those marriages that took place in 1987, 58 per cent of married men and 53 per cent of married women said that they had lived with their future spouse before marriage. Most pre-marital cohabitation was only for short periods with 40 per cent cohabiting for less than a year. The highest proportions of those cohabiting were in their 20s and early 30s, with higher proportions of older men (over 45) cohabiting than older women. Regional variations were also evident with the highest proportions occurring in East Anglia and the south east and the lowest in Scotland and Wales (Haskey and Kiernan, 1989).

Cohabitation has been reported to be increasing in Luxembourg (Als, 1990) and the Netherlands where there is increased acceptance of modern household forms and traditional opinions about marriage show no signs of returning. In the Netherlands between 1979 and 1982 the percentage of women aged 20–4 living in consensual unions rose from 10 to 16; and in a 1984 survey, two-thirds of cohabiting couples said the birth of a child would be a reason for them to marry (Keilman, 1987).

Data on cohabitation are rather poorer for most other EC countries. No statistics are collected in Belgium, although one piece of research showed that 10 per cent of all women studied and 40 per cent of divorced women had cohabited at some time (Sels and Dumon, 1990). In Italy it appears that cohabitation is a much more limited phenomenon. A survey by the National Statistical Institute in 1983 found 192,000 cohabiting couples, or 1.3 per cent of all couples, and only 3 per cent of these were women aged under 24. The majority of the consensual unions were formed by older people who did not wish to marry for fear of losing a pension from an earlier marriage. Cohabitation also appears to be geographically concentrated in the large cities of north-west Italy (Golini, 1987). Under-reporting is likely and the true picture may be somewhat higher. Cohabitation is still low in Spain, although the recent increase in extra-marital births (from only 2.8 per cent of all births in 1979) may be associated with an increase in the proportion of cohabiting couples (Del Campo, 1990). There are no data on cohabitation in Greece, although the situation is likely to be similar to the one prevailing in Italy and Spain where there is still considerable social pressure for legal marriage (Symeonidou, 1990). In Portugal, where the current percentage of extra-marital births (13.7 in 1987) is much higher than in other southern European countries, one particular area south of the River Tejo has a certain tradition for consensual unions (Hopflinger, 1985).

The effects of the rise of cohabitation on family values may be relatively weak, although it is certainly part of the process making the family a more open system and clearly it has reinforced the decline of the institution of marriage. The increase in cohabitation is a response to a number of changes including women's greater equality in society. Cohabitation may also be seen as a more effective way of creating an egalitarian power structure within a relationship, one that reflects women's greater autonomy (Hopflinger, 1985). Non-marital cohabitation has emerged in the 1980s as a concomitant to declining marriage rates. It may be seen both as a response to disillusionment with marriage and as a strategy to reduce the risk of divorce. Cohabitation has also given greater fluidity to relationships which may, of course, be formed and broken more easily than legal marriages.

### *Divorce*

The increase in the chance of a marriage ending in divorce is perhaps the single most important change affecting the family in Europe today. Marital breakdown has many repercussions; in particular, it has a dramatic impact on the lives of the couples concerned as well as on their children. Increasing divorce rates are also largely responsible for the growing number of one-parent families. Remarriage may also be more common in which case the numbers of stepparents and stepchildren are bound to rise along with increasingly complex family relationships.

The number of divorces in the 12 countries of the Community rocketed between 1960 and 1988, increasing from about 125,000 to 534,000. Overall, the divorce rate tripled between 1964 and 1982 (figure 9.2). Variations between countries reflect the current state of legislation governing divorce so that rates are still very low in southern Europe, where laws are restrictive, and non-existent in Eire, where divorce is prohibited. Elsewhere, divorce rates were increasing in the 1960s prior to the general relaxation

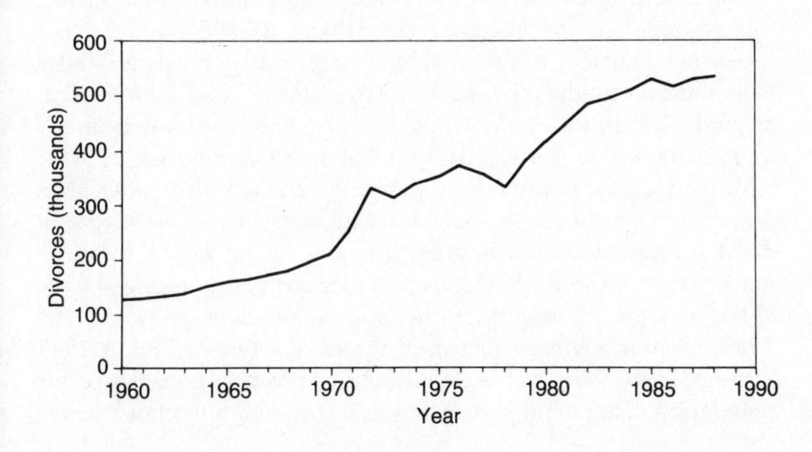


**Figure 9.2** Trend in the number of divorces in the European Community.

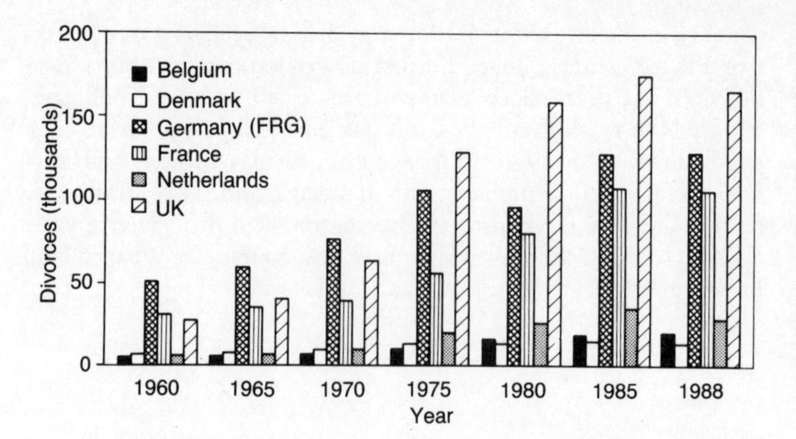


**Figure 9.3** Trends in the number of divorces in selected European Community countries.

of divorce laws around 1970. In subsequent years rates increased rapidly as divorce became more frequent and occurred earlier in marriage.

The index of divorce, which takes duration of marriage into account, increased between two and four times in different member states of the Community between 1965 and 1980. The percentage of marriages that began in 1975 and, it is estimated, will finally end in divorce is 21 in France, 27 in England and Wales and 32 in Denmark (Roussel, 1987). For marriages that began in 1987 in England and Wales and are experiencing current divorce rates, it has been estimated that 37 per cent will end in divorce (Haskey, 1989a). Current rates of divorce in northern EC countries range from 8 per thousand marriages in Belgium, the Federal Republic of Germany and the Netherlands, to 12.3 and 13.1 in the UK and Denmark, respectively (EUROSTAT, 1990; see figure 9.3).

The rise in the divorce rate began to slow down in the 1980s; for example, in the UK the rate peaked in 1985 at 12.9 per thousand married couples; it had previously risen following amendments to the divorce law. By 1988 it had stabilized at around 12.3. In the Netherlands there has been a decline from 9.9 in 1985 to 8.1 in 1988. The apparent stabilization may coincide with the increase in preference shown for cohabitation rather than legal marriage. In those countries with very low or non-existent divorce rates, desertion is often adopted as the means to end a marriage. In Italy, while the divorce rate has hardly changed from about 0.3 per thousand of the population between 1971 and 1986, separations have increased threefold from 11,796 in 1971 to 34,332 in 1986, that is, from 0.22 to 0.6 per thousand of the population. This trend is reflected in one Italian household survey which shows that 1.4 per cent of married couples did not live together (Golini, 1987).

Fewer divorced people are remarrying than in even the recent past. In 1977 in France about 60 per cent of divorced people remarried; by 1981 the figure had fallen to about 46 per cent (Roussel, 1987). Increasingly, cohabitation is favoured to remarriage so that in the Scandinavian countries non-marital cohabitation is by far the most frequent form of permanent sexual union.

Second marriages are even more likely to end in divorce than first marriages. It has been estimated that in England and Wales the marriage of an already divorced man is one and a half times more likely to end in divorce than that of a man marrying at the same age for the first time, and a divorced woman is twice as likely to divorce again than her never-married counterpart (Haskey, 1983). And since increasing proportions of marriages involve at least one divorced partner – in England and Wales about a third of marriages from 1981 onwards compared with only a sixth in 1971 – so marriages become even more fragile.

The impact of increasing divorce rates has been considerable. The disruption to family life has an impact on all involved, but especially the children. In England and Wales in 1961 there were 35,000 children with divorced parents; this increased to 81,000 in 1971 and 149,000 in 1987 (Haskey, 1988a).

There have been many attempts to account for the rise in divorce rates, for example, in terms of easier divorce laws, the increasing independence of women and changing family values. Hopflinger (1991) has argued that the reasons are probably to be found in changing socio-cultural values; marriage has become an institution for companionship based on love rather than one intended to ensure economic and reproductive survival. Thus marriage is increasingly dependent on the maintenance of a strong emotional attachment which may make its breakup more likely (Roussel and Festy, 1979). It should also be remembered that in the past marital dissolution was also high because mortality was high and poverty was often accompanied by desertion. By comparison, the relative stability of marriages between the Second World War and the early 1960s can now be seen to be exceptional (Hopflinger, 1991).

### *One-parent Families*

The increasing incidence of divorce since the 1970s has been accompanied by an increase in the number of one-parent families. In Europe today between 6 and 12 per cent of all family households are headed by a lone parent who is most likely to be a

**Table 9.4** One-parent families as a percentage of all family households around 1981

| | *Lone mother* | *Lone father* | *Total* |
|---|---|---|---|
| Belgium | 7.7 | 1.9 | 9.6 |
| Denmark | 9.9 | 1.5 | 11.4 |
| Eire | 10.1 | 2.6 | 12.6 |
| France[a] | 5.2 | 0.9 | 6.1 |
| Germany (FRG) | 7.9 | 1.5 | 9.4 |
| Italy | 7.5 | 3.0 | 10.5 |
| Luxembourg[b] | 7.7 | 1.6 | 9.3 |
| Netherlands | 6.8 | 1.3 | 8.1 |
| Spain | 6.7 | 1.5 | 8.2 |
| UK | 9.0 | 2.5 | 11.5 |

[a] 1982.
[b] 1980.
*Source*: Hopflinger, 1991, table 10, p. 35

divorced women, but may also be widowed, separated or single (table 9.4). Lone parent families take a variety of forms including some that are legally classified as lone parent even though co-habitation is involved. A survey in the Federal Republic of Germany showed that 10 per cent of formally lone mothers were in fact cohabiting, as were 28 per cent of lone fathers (Hopflinger, 1991), and equivalent data on one-parent families are by no means straightforward.

Although one-parent families were relatively numerous in the 1960s, they were typically the result of widowhood. Again in the Federal Republic of Germany, 10.5 per cent of families were headed by one parent in 1961 and of those headed by a woman 46 per cent were widows and 24 per cent were divorced. The percentage of one-parent families declined to 8.7 in 1970 and then rose to 11.4 in 1982, by which time only 24 per cent of the lone mothers were widows while 45 per cent were divorced (Schwarz, 1986). The pattern was similar in the Netherlands where the percentage of one-parent families remained constant at 10 between 1971 and 1983. In 1971, 65 per cent of lone mothers were widows and only 17 per cent were divorced; by 1983 21 per cent were widows while 60 per cent were divorced (Clason, 1986). Similarly, in France the overall increase in one-parent families was small, rising from 8.7 per cent of all families with children in 1968 to 9.8 per cent in 1982, but once again the main cause of lone parenthood had changed from widowhood to divorce. In 1968, 54 per cent of lone parents were widowed and only 17 per cent were divorced; by 1982 31 per cent were widowed and 39 per cent were divorced (Lefaucheur, 1986). In Great Britain, the percentage of one-parent families headed by a divorced woman increased by 15 between 1972–4 and 1982–4 while those headed by widows declined by 9 per cent (Haskey, 1986). By contrast, in southern Europe, where divorce rates are still very low, the majority of lone parents are widowed. In Italy, for example, where 14 per cent of families with children were headed by a lone parent in 1981, 73 per cent were widowed (Golini, 1987).

The other category of lone parent which increased particularly from the 1970s onwards is that of single mothers. This reflects the increase in extra-marital births in many countries of the EC together with the increasing acceptability of motherhood among single women. In Great Britain, one-parent households headed by single mothers increased by 6 per cent between 1972–4 and 1982–4. In the Netherlands, 3 per cent of lone mothers were single in 1960 and by 1983 the figure had increased to 13 per cent (Clason, 1986).

The increase in divorce, along with extra-marital births and single non-cohabiting mothers, has led to an increasing number of children living with a lone mother. In Great Britain, for example, while 8 per cent of all children were living in one-parent families in 1972 (7 per cent with a lone mother and 1 per cent with a lone father), by 1986 12 per cent were living with a lone mother and 2 per cent with a lone father (Haskey, 1989b). Similarly in the Federal Republic of Germany, while 6 per cent of children were living with a lone mother and 1 per cent with a lone father in 1970, by 1985 9 per cent were living with a lone mother and 2 per cent with a lone father (Hopflinger, 1991).

But the picture of one-parent families remains an obscure one. One-parent families may be only a temporary state before the remarriage or cohabitation of the lone parent. A British study in 1980 showed that the median duration of lone parenthood was three years for single women compared with five years for divorced women, although the mean duration for all divorced women was about 7.5 years. The length of time alone varied by a woman's age and the number of her children: older women and women with larger numbers of children spend longer periods in lone parenthood (Ermisch, 1989). With remarriage, children from one-parent households move into so-called 'blended families', with a step-parent and perhaps step-siblings and half-siblings. But they do not lose their other parent, so that the family network of such children becomes highly complex with divergence between their family household and their wider family. In the Federal Republic of

Germany in 1982, 7.5 per cent of children were living in a step-family (Schwarz, 1986), while in Great Britain at the same time 3 per cent of the under five age-group, 5 per cent of the 5–9 age-group and 8 per cent of the 10–15 age-group were living in a step-family.

The economic position of many one-parent families is difficult, particularly if headed by a divorced woman. In Great Britain in 1984, 65 per cent of one-parent families were receiving social security supplementary benefit (Haskey, 1986), and even after adjusting for household composition their average income was only 60 per cent of that of households with children and two parents (Ermisch, 1989). A Dutch survey of 1983 showed that 75 per cent of divorced women with younger children lived on an income at the social minimum (van den Akker and van der Avort, 1986). If families with children are becoming increasingly disadvantaged in societies where childless families and households are the norm, then one-parent families are likely to be even more disadvantaged. Yet increasing proportions of children will at some time in their early years be part of such a family: current levels of divorce in Great Britain suggest that one in five of all children will experience the divorce of their parents before they reach 16 (Haskey, 1986).

### *The Decline of Fertility*

The decline of fertility which has occurred throughout the EC has had a dramatic impact on the family and household (see chapter 5). The most immediate ways are through smaller family size and the reduction in the number of households with children. By 1980 less than half of all households had children in them so that increasingly the typical European household is no longer made up of wife, husband and children but is more likely to be childless (table 9.5). In Great Britain, the percentage of households with dependent children declined from 40 in 1961 to 33 in 1984, while no-family households increased from 17 to 28 per cent in the same period (Haskey, 1986). Even in Italy in 1981, the classic family of parents and children represented only 53 per cent of all households, although 67 per cent of the population did live in such households (Golini, 1987).

Not only are there fewer households with children, but there are fewer children in households as a result of the reduction in third and higher order births since 1960 (table 9.6). Apart from Spain, Portugal and Eire, where just over half of the adults live in households with children, the majority of those aged over 15 in all EC countries now live in households without children. Only in

**Table 9.5** Percentage of households with children in selected European Community countries

| | *Around 1970* | *Around 1980* |
|---|---|---|
| Eire | | 60.4 |
| France (0–16) | 39.6 | 35.7 |
| Great Britain (dependent children) | 38 | 36 |
| Italy (all unmarried children) | | 59.3 |
| Netherlands (<18) | | 49.3 |
| Germany (FRG) (<18) | | 32.1 |

*Source*: Hall, 1988, p. 22

**Table 9.6** Third or higher order births as a percentage of total births, 1960 and 1988

| *Country* | *1960* | *1988* |
|---|---|---|
| Belgium | 37.3 | 18.9[a] |
| Denmark | 36.0 | 16.3 |
| Eire | 60.9 | 41.1 |
| France | 38.8 | 25.0 |
| Germany (FRG) | 28.2 | 16.9 |
| Greece | 27.4 | 16.2 |
| Italy | 34.8 | 16.6[b] |
| Luxembourg | 27.8 | 17.1 |
| Netherlands | 41.8 | 20.5 |
| Portugal | 45.3 | 18.7 |
| Spain | 32.0[c] | 21.6[c] |
| UK | 33.2 | 23.8 |

[a] 1986.
[b] 1987.
[c] 1975 and 1985.
*Source*: EUROSTAT, 1990

Eire does a significant percentage (9.5) of adults live in households with three or more children (Hopflinger, 1991).

The other aspect of the prevailing fertility pattern that has had an impact on families has been the increase in extra-marital births during the 1980s. This has been especially important in Denmark, France and the UK, but elsewhere the increases have been smaller. In many EC countries about 10 per cent of births now occur outside marriage (table 9.7).

**Table 9.7** The percentage of live births outside marriage, 1960 and 1988

| *Country* | *1960* | *1988* |
|---|---|---|
| Belgium | 2.1 | 7.9[a] |
| Denmark | 7.8 | 44.7 |
| Eire | 1.6 | 11.7 |
| France | 6.1 | 26.3 |
| Germany (FRG) | 6.3 | 10.0 |
| Greece | 1.2 | 2.1 |
| Italy | 2.4 | 5.8 |
| Luxembourg | 3.2 | 12.1 |
| Netherlands | 1.4 | 10.2 |
| Portugal | 9.5 | 13.7 |
| Spain | 2.3 | 8.0[b] |
| UK | 5.2 | 25.1 |

[a] 1986.
[b] 1985.
*Source*: EUROSTAT, 1990

### *Women*

The changing role of women in society and within the family has occupied a central place in the transformation of the family and household since the 1960s. These changes have been described as a move from the housewife to the employed woman pattern (Jallinoja, 1989). Closely related are the decline in fertility from the mid-1960s and wider changes in the attitudes and expectations of women which are reflected by rising rates of divorce and co-habitation and declining marriage rates. Family life has changed in roughly the same manner in all the EC countries, although there have been variations in the speed of change. Relatively little happened in the 1950s when the family was demographically and socially stable. The 1960s was a period of transition in which the decline of fertility and the emergence of the women's movement coincided. Changes were much more rapid in the 1970s with the development of a diversified and less stable family model (Dahlstrom, 1989). This process has continued in the 1980s and 1990s. It can be argued that the changing role of women in society is central to these processes: the women who have participated in these changes were born after the Second World War, are better educated and perceive wider opportunities for themselves than women in earlier generations.

The increased participation of women in the paid labour force has been part of the process, although participation rates continue to vary from country to country, especially among women with young children (see chapter 13). Many women work part-time as opposed to full-time in response to the problem of combining family responsibilities with paid work. For many women the conflicting demands of reproduction and childcare and paid work persist since women continue to have the primary responsibility for managing the family and children, irrespective of who may take care of the children while they are at work, and for routine domestic work. Men continue to participate far less in such work. While the family is a restraining influence on women in paid work, paid work strongly limits men's participation in family life (Haavio-Mannila, 1989).

Among EC countries, Denmark, the UK and France have higher female participation rates in paid employment, especially at older ages. By contrast, many women in Eire, Belgium and Spain stop work for child-bearing and do not resume it thereafter. The Netherlands, Italy and the Federal Republic of Germany occupy an intermediate position (table 9.8).

In Great Britain there has been a rapid increase in the number of women with paid work who also care for pre-school-age children. In 1985, 27 per cent of women with one child worked, often on a part-time basis, and in 1989 it was 39 per cent. Indeed, the average number of hours a week worked by British women was

**Table 9.8** Employed women as a percentage of all women in the age-group

| | | *Age-group* | | | | | |
|---|---|---|---|---|---|---|---|
| | | *20–4* | *25–9* | *30–4* | *35–9* | *40–4* | *45–9* |
| Belgium | 1981 | 71 | 73 | 63 | 55 | 45 | 38 |
| Denmark | 1986 | 83 | 87 | 89 | 88 | 87 | 82 |
| Eire | 1987 | 75 | 59 | 40 | 29 | 28 | 28 |
| France | 1987 | 64 | 76 | 72 | 72 | 72 | 68 |
| Germany (FRG) | 1986 | 74 | 67 | 62 | 62 | 62 | 58 |
| Great Britain | 1986 | 69 | — 63 — | | — 71 — | | |
| Greece | 1985 | 46 | 51 | 51 | 52 | 47 | 45 |
| Italy | 1987 | 63 | 62 | 60 | 60 | 56 | 49 |
| Luxembourg | 1987 | 72 | 65 | 54 | 49 | 42 | 36 |
| Netherlands | 1987 | 74 | 65 | 54 | 56 | 54 | 48 |
| Portugal | 1987 | 66 | 73 | 74 | 69 | 62 | 56 |
| Spain | 1987 | 60 | 60 | 49 | 38 | 35 | 31 |

*Source*: Hopflinger, 1991, table 6

30, rather less than among women anywhere else in the Community (*Social Trends*, 1991). In Denmark changes have been even more dramatic: while in 1974 43 per cent of women with children aged 0–6 were described as housewives and 29 per cent worked for more than 25 hours a week, in 1985 only 8 per cent were housewives and 59 per cent worked more than 25 hours a week, with 35 per cent working at least 40 hours. In 40 per cent of Danish families with children aged 0–11 both father and mother work at least 38 hours a week (Soren, 1990).

If anything, the changes since the 1960s have brought increasing conflict for women between paid and unpaid work as well as rapidly increasing workloads. With such pressures it is not surprising that childlessness or one-child families are increasing. For example, in France, of those women born in 1955, 11 per cent were still childless in the mid-1980s, while in England and Wales and the Netherlands the figures were 18 and 20 per cent, respectively. For couples married between 1970 and 1974 in the Federal Republic of Germany, 19 per cent were still childless by 1988 (Hopflinger, 1991). Certainly the double burden of paid and unpaid work will continue to affect reproductive behaviour, and the perpetuation of the small family seems inevitable.

### *One-person Households: the Young and the Elderly*

The greater diversity of family and household forms is well illustrated by the dramatic increase of one-person households (table

**Table 9.9** One-person households as a percentage of all households

| | *1960* | *1970* | *1980* | *One-person elderly households* |
|---|---|---|---|---|
| Belgium | 17 | 19 | 23 | |
| Denmark | 20 | 21 | 29 | |
| Eire | 13 | 14 | 17 | 7 |
| France | 20 | 22 | 25 | 12 |
| Germany (FRG) | 21 | 25 | 31 | 14 |
| Great Britain | 15 | 18 | 22 | 14 |
| Greece | 10 | 11 | 15 | |
| Italy | 11 | 13 | 18 | 11.5 |
| Luxembourg | 12 | 16 | 21 | |
| Netherlands | 12 | 17 | 22 | 9 |
| Portugal | 11 | 10 | 13 | |
| Spain | | 8 | 10 | |

*Sources*: Hall, 1988; Keilman, 1987

9.9). This results from the increasing proportion of the elderly in the population, together with the sex-specific mortality differential which leads to an ever increasing number of elderly widows (see chapter 8). Alongside the rise of the one-person household there has been a decline in the number of co-resident three-generation families. The elderly are more likely to live alone and not with their children or other relatives than in the past, and particularly in rural societies. The number of three- and four-generation households declined by about 50 per cent in the Federal Republic of Germany between 1961 and 1982, that is, from 5 to 2 per cent of all households (Schwarz, 1988). Such multi-generational households have persisted longer in southern Europe, although the 1981 Italian census spoke of the 'decomposition of the extended family with the phenomenon of the isolation of the elderly' (Hall, 1988).

There has also been an increase in the number of adults living alone, especially in those countries with high rates of cohabitation. The one-person household is often a transitional state for young adults as they move from the parental household to cohabitation. Marital breakdown has also contributed to the increase in younger adults living alone.

In the past, marriage was often the point at which young adults left their parental home, and even in the 1960s there was a close coincidence between the two events. Since then the link has weakened. In 1982 a survey of young people in the EC showed that among those aged 20–4 the percentage of men living alone varied from 29 in Denmark and 21 in the Federal Republic of Germany to 4 per cent in Eire and the UK. For women, percentages varied from 29 in Denmark and 21 in the Federal Republic of Germany to 6 and 3 per cent in the UK and Eire, respectively. To some extent the variations reflect housing market differences between the countries. For example, in Denmark there is a greater supply of reasonably priced accommodation for rent than in the UK (Kiernan, 1986).

## Policy Issues

It is clear that future changes in Europe's population will stem from developments in the structure and organization of the family, but particularly from how the tensions between women's various roles and their involvement in the family are to be resolved.

There is a variety of policy responses to the family in Europe. Some countries are still adapting to the decline in importance of the traditional family and the emergence of much more diverse family structures. Others are concerned with the tensions that exist between home and paid labour, especially for women.

Only a few countries have an explicitly stated family policy; notably France, Belgium, Luxembourg and the Federal Republic of Germany. Apart from the last mentioned, policies towards the family in these countries are pro-natalist and emphasize the need to create a climate of opinion more friendly to children. In France the family has been central to government policy, more so than in most other EC countries. Following the Second World War family policy was constructed in order to give preferential treatment to a specific model of the family, one with at least three children and a mother at home. In 1959 family benefits represented 28 per cent of all social security benefits; now they represent only 11 per cent. In fact the French government is now much more neutral towards the family and recognizes the fuller emancipation of women and the need to respect individual choice. There is also more sympathy for those attempting to reconcile the demands of family and professional life (Commaille, 1990).

In other countries, such as the Netherlands, Denmark and the UK, the policy is implicit and comes under the broader heading of social policy. Even so, the emphasis in these countries varies. Denmark, where changes in family structures have in many ways gone furthest, has the more developed policies towards families with children. The focus is specifically on the needs of children whatever type of family they live in (Soren, 1990). In the UK there is a variety of interest groups concerned with the family, but there is also increasing divergence of opinion on matters of principle which tends to discourage the formulation of a coherent family policy (Wicks and Chester, 1990). In the Netherlands, on the other hand, there are few specific interest groups whose principal concern is the family, and although many different aspects of government policy affect the family there is still a strong emphasis on retaining its autonomy (Jonker, 1990).

Family policy is much less well developed in Spain, Portugal and Greece, although there is growing awareness of their particular needs and those of children. In Italy it has been argued that the lack of government support has contributed to the fragmentation of the family and the decline of fertility, but even so the family still holds a symbolically important position in Italian culture (Donati, 1990). Similarly, in Eire the family is regarded as the fundamental unit of society, yet there is no co-ordinated social policy towards the family nor is there an awareness of the growing problems caused by the deterioration of family incomes in relation to other incomes (O'Higgins, 1990).

The 1975 Greek constitution states that the family is the foundation stone for the maintenance and promotion of the nation, but even so there is limited special assistance for families. Child and family allowances are very complex and time consuming procedures are necessary to obtain them. Even though female participation in the labour force is lower than elsewhere in the Community, changes are taking place. In Greater Athens, for example, 33 per cent of married women with children aged under 6 are working and thereby generating a great demand for child-care (Symeonidiou, 1990).

Throughout the EC the issue of the family is surrounded by political rhetoric. A debate has developed on the tensions within the modern family, particularly the conflict between paid and unpaid work and its impact on women and thence on the family itself. Recognition is growing of the need to support the family in its important child-rearing function and that childcare is a parental and not just a maternal responsibility. However, in most countries there is still more rhetoric than action over the need to create working conditions for women and men that will make it easier to combine the parental roles with paid work. There is also the danger that as the proportion of families with children declines so the needs of that section of the population currently engaged in child rearing will be politically marginalized.

There are several other issues that relate to the family and social policy. For example, the so-called crisis of the welfare state which faces several EC countries has given renewed impetus to the creation of informal caring networks. Any shift from institutions to the local community generally implies care within the family or by informal caring networks which are usually dependent on women. The growing number of elderly in the population also require caring networks and, although the elderly may no longer live with their children, contacts are still important, especially because it is often via family contacts that old people strive to maintain their independence. In general there is a growing dissociation between family kin networks and households, not just among the elderly but also as a consequence of marital breakdown and household dissolution.

## Concluding Remarks

Households and families have changed rapidly over the last 30 years and these changes are likely to be maintained in the near future. It is still debatable whether a measure of stability has been reached in some EC countries in relation to divorce, child-bearing or cohabitation, and how far the experience of the countries of southern Europe will follow that of the northern countries. European society will have to adjust to a variety of household and family forms and to the tendency for more transitional household types to increase in number as many more individuals pass through more complicated life-cycle stages. Households and families are moving inexorably towards smaller units as the result of a number of factors: ageing, declining fertility, increasing independence and female emancipation included. But it also seems unlikely that a single uniform European household type will emerge. For, while in many respects the traditional child-centred nuclear family may be seen to be weakening, the responsibilities of the family are not diminishing.

The fundamental changes taking place are not a result of political or economic factors peculiar to a particular region or social background. Major transformations are under way in all Western, post-industrial societies, including the countries of the EC. The future of the family is uncertain, although some form of social arrangement involving a permanent sexual union in which reproduction and childcare continue to be important functions will persist, with or without marriage and divorce.

# 10

# Education

Jean Renard

Every educational system seeks to achieve two objectives. First, it aims to widen students' knowledge, to develop their faculties of comprehension and intervention in their environment, to reduce ignorance and to distribute knowledge and power more equally. Its second goal is to train a qualified, active population to adapt themselves to the needs of the economy and industry. How do these social and political ambitions translate statistically within the European Community (EC)? How has education evolved in Europe and what form does its contemporary structure take? Setting aside the intentions of individual countries, can convergent trends in educational policies be observed in the Community?

## The Evolution and Structure of Europe's Student Population

The number of full-time pupils in Europe increased rapidly in the 1960s and 1970s, but has declined continuously during the last ten years. (The definition of full-time pupils used here excludes alternative post-school professional education which is widespread in some countries, such as Germany.) Estimated at 63 million, or 21 per cent of the total population, at the beginning of the 1970s, the number of full-time school pupils reached a maximum during the year 1977–8 of nearly 72 million, or just over 23 per cent of the population. Since 1977–8 census figures show a constant decrease to around 68 million in 1986–7, again 21 per cent of the total population. In certain years the decline was in excess of 1 per cent. According to the EC's Statistical Office, this downward trend should continue until at least the end of the century.

The main cause of the decline has been the fall in the birth rate. Since the mid-1960s, which marks the end of the baby boom, the annual number of births in the Community has been reduced by a third. (See chapter 5 for a fuller account of European fertility trends.) The effects of this impressive drop in the number of births over the last 25 years or so has been partly counter-balanced by the increase in participation rates, especially in the under-6 and over-14 age-groups. The number of older secondary and higher education students has recorded the most positive development and it is the increase in this sector that is responsible for offsetting what would otherwise be an even more dramatic general reduction in the number of European students. From the mid-1970s to the mid-1980s, the number in this category increased by about 4 million, an increase of about 33 per cent, from 14.3 to 19.2 million, whilst during the same period the size of the total school population at all levels was reduced by about 4 million.

The number of pupils attending primary or lower secondary school levels has diminished noticeably as a result of the fall in the birth rate. Even if there was an eventual rise in school participation rates, the number of pupils would not increase since schooling is compulsory at this age. The number in this category fell over ten years from around 48 million to a little more than 40 million, a decrease of over 15 per cent. In contrast, after a period of increase due to the aftereffects of the baby boom and to changes in parental attitudes, the size of the pre-school population has stayed largely the same since the mid-1970s at between 8.5 and 8.6 million. The effects of the fall in the birth rate have largely been offset by the increase in the popularity of pre-school education and thus an increase in the participation rate which is approaching 100 per cent among 4 year olds in some European counties.

In France the only change that is likely to occur in the near future will involve an increase in the number taking advantage of early pre-school education for those under 3 years old. National policies on nursery education vary widely. In Great Britain, for example, the number of pre-primary-school children still remains at about 400,000, while in France, by contrast, the number currently stands at between 2.5 and 2.7 million.

The developments described above concern the EC as a whole. However, although convergence of trends is obviously continuing, there are still some significant differences to be observed. How may we classify the 12 countries of the Community in terms of the evolution of their school populations? To answer this question we must examine variations in total school attendances divided up into different levels of study over more than ten school years.

Four countries, each one of which experienced a substantial increase in the number of pupils between 1975 and 1987, can be grouped together: Greece, with a 12 per cent increase; Eire with 14 per cent; Portugal with 20 per cent; and Spain with 22 per cent. This general increase may be explained by several factors: recent demographic history, the late decline of the birth rate or its persistence at a high level and the drive to extend the education system. During the 1980s the struggle against illiteracy was still being pursued with vigour in Greece, Spain and especially Portugal where more than 20 per cent of those aged 15 and over were recognized as illiterate in 1981. This is why the EC Commission responsible for matters of employment, social welfare and education published a report in 1988 dealing with the struggle against illiteracy. In September 1988 the Ministry of Education in Portugal published the results of a study entitled Programme for Educational Development in Portugal for the period 1988–9 to 1992–3. This report highlights the educational backwardness of Portugal, compared with the Community at large, as well as the deficiencies of the national educational system, the significance of illiteracy, the inadequate provision of pre-school education which is still in its infancy and the reduced number of students in secondary and higher education.

Five countries form a second group: Denmark, Germany, Italy, the Netherlands and the UK. They are characterized by both an early decline in the birth rate and a high standard of living; all have experienced a reduction of their total school population. But while the numbers in primary and secondary education in particular have fallen, those in the over-16 category have increased.

France seems to bridge the gap between these two groups of countries. The total school population rose slightly over the ten-year period. The increase in the rate of schooling for the under-6s and over-16s appears to have been sufficient to compensate for the effects of fertility decline which was anyway less marked in France than in other European countries.

Two other countries, Belgium and Luxembourg, should be treated separately. Like their neighbours they recorded a reduction in the total number of pupils, but in Belgium official statistics since 1981–2 do not distinguish between higher and lower secondary school pupils. In Luxembourg, on top of the statistical effects resulting from the modest number of students (about 60,000 in the early 1980s), there is also a repeated absence of data on pre-primary and primary school aged children. However, taking into account their recent demographic history, it seems reasonable to place Luxembourg in the second group of countries.

In 1975 the percentage of 19–24 year olds in full-time education varied considerably among European countries. While the European average for male students was about 20 per cent, in Denmark it was 27 per cent and in Germany, Italy and the Netherlands it was 23–5 per cent. On the other hand, it was less than 14 per cent in Eire and Portugal. For female students the figures varied from 25 per cent for Denmark to 8–9 per cent for Eire, Portugal and the UK. Apart from Italy, the other Mediterranean countries recruited few young women into higher education.

Although the numbers of both male and female students have generally increased, we must not overlook the fact that differences between European countries had widened by 1986–7. Thus, for men and women in the 19–24 age-group the percentage of students remained small at about 10 per cent in Portugal whereas in Belgium, Germany, France and Spain it crossed the 25 per cent threshold.

The stable, but low, figures for students in the UK are surprising (12 and 9–10 per cent for males and females, respectively). In the mid-1980s, less than 590,000 students were recorded in this age-group in the UK, compared with 1 million in France, 1.1 million in Italy and more than 1.3 million in West Germany. The length of courses and differences in the requirements for entry to higher education explain some of these differences. While in France, a baccalaureat pass opens a direct door to higher education, in the UK and Eire the right to further study in a higher education institution is not automatic even if the student has attained the necessary minimum qualifications. Access depends on selection and individual institutions have the final say. In Spain, candidates must take an entrance examination to get into higher education. Nevertheless, recruitment remains fairly open with 860,000 students in 1986.

There have been particularly significant developments in the numbers of women students in upper secondary and higher education since the early 1970s. For both of these levels the increase in the number of women has been more rapid than that for men. In the early 1970s women represented 46.2 per cent of the total number of students enroled in upper secondary education and less than 37 per cent of the total enrolled in higher education. Ten years later the percentages reached more than 50 and over 44, respectively. The rate of schooling by gender in the 15-plus age-groups has therefore increased to different degrees. The number of women in higher education rose from 14 to 19 per cent between 1975 and 1985 and from 48 to 63 per cent in upper secondary education. For men the rates increased from 20 to 21 per cent and from 50 to 59 per cent, respectively. Although the progressive 'feminization' of the mostly post-16 student population is occurring throughout Europe, it is doing so at different rates in different countries. While the change remains modest in some north-west European countries – France, Germany, Denmark and the UK – there has been an enormous increase in the Mediterranean countries. This is without a doubt the most important development observed in the field of education in the Community in recent decades. Enumerated at 27,000 in the early 1970s, the female student population of Greece increased to over 80,000 by the mid-1980s. In Spain the increase was equally remarkable for the same period, the numbers in upper secondary education increasing sixfold and those in higher education, fivefold.

This 'feminization' of the post-16 student population reveals, and at the same time provokes, profound socio-economic upheaval in the Mediterranean parts of Europe. The change has been accompanied by a very rapid decline of fertility. It expresses an important modification to the position of women in society and challenges traditional assumptions about gender-specific roles in employment and the family (see chapters 9 and 12).

The distribution of students by subject studied shows distinctive patterns. Four categories of subjects cover just over two-thirds of all EC students. The four categories in order of popularity are social studies, medicine and hygiene, engineering and industrial training, and natural sciences, mathematics and computer studies.

Of all the EC's students, 26 per cent are currently involved in social studies courses which include such diverse subjects as the behavioural sciences, commercial and administrative training, domestic science, subjects leading to careers in the tertiary sector, and computing and documentation management. The attraction of these different subjects underlines the dominant role of the tertiary sector in our economy and society. On its path to a post-industrial economy, European society must reinforce the educational and professional interests of information and communication, which will doubtless prove to be the sources of progress in the next century.

The medical sciences of health and hygiene attract 14 per cent of EC students with young women comprising over 60 per cent. However, the proportion of female students varies inversely with the length of professional training. According to the International Classification of Education which distinguishes a category of relatively short training which could be compared with the first one or two years of higher education and a category of advanced training which corresponds to longer studies leading to the doctorate level, in 1986–7 the percentage of female students was 85 in the first category but only 48 in the second. The importance of this medical domain and of health in general in the educational system reflects the attention given to this subject by Europeans. It is a reminder of the pertinence of the present debate over the rapid increase in public expenditure in these health-related fields by countries with ageing populations. Nevertheless, although medical sciences attract more than 15 per cent of students in Greece, Italy and France, to the extent that a very selective system of entry examinations is deemed necessary, in other Community countries the proportion is much lower, under one-tenth in Eire and Portugal for example.

Engineering sciences come next with a little less than 13 per cent of European students in higher education. This includes training for industrial production and associated activities, as well as for transport and telecommunications. The percentage of young women is small in these subjects, varying from 7 to 14 per cent depending on the type of training involved. The significance of this sector of education reflects the need for advanced technical training as well as for continuous technological innovation in order for European companies to maintain their competitive positions against Japanese and American competition. But the European average conceals important national disparities which might help to explain differences in the efficiency of industrial enterprises. Engineering and related studies attract more than 17 per cent of students in Germany and the Netherlands, whereas in France, Italy and Spain the figure is below 10 per cent. In the Netherlands and Denmark the supply of engineers exceeded demand in 1990, whereas there was a deficit in the UK, France and even Italy.

Finally, the natural sciences, mathematics and computing in combination represent 11–12 per cent of EC students. The attention given to these scientific subjects by different countries reveals the demand for scientists, computer operators and various industrial specialists based in laboratories. There are sharp differences in emphasis, however. For example, in Italy there has been particularly poor representation in the area of vocational training; the country has even been described as one populated by jurists and literary experts.

### From School to Work: Problems of Interaction and Adaptation

The second fundamental objective of any education system is to train a qualified working population capable of adapting to the needs of industry. However, since the 1970s the large increase in the number of unemployed young people has revealed a certain mismatch between the educational system and the demands of the professional world. Social interaction and involvement for young people depends on their initial involvement in the workforce. But for a large number of under 25 year olds, introduction to the workforce takes place under difficult conditions such as high levels of long-term unemployment, unstable or part-time jobs, the devaluation of qualifications, fixed-term contracts followed by periods of unemployment and repeated participation in courses to prepare for professional life.

During the mid-1980s more than one under 25 year old in five was unemployed compared with one in ten for the entire economically active population of the Community. Moreover, the situation varied considerably from one member country to another. In Denmark, Germany and Luxembourg the youth unemployment rate was less than 14 per cent, but in the Netherlands, Great Britain, France and Greece it varied between 17 and 24 per cent and in Italy it was nearly 35 per cent (see chapters 11 and 12).

The key to success in introducing young people into the workforce relates to the level of qualifications and the quality of professional training offered as well as to the training and preparation of young people provided by public and government agencies. Much of this training or work experience is conducted beyond the traditional forms of schooling. These developments are altering the nature of education throughout the Community.

In order to face the increase in youth unemployment in the 1970s and 1980s, the EC countries developed a wide range of educational, financial and economic strategies. Germany enlarged her old system, the dual system of alternating professional teaching and apprenticeship. After compulsory schooling, training is organized in conjunction with a professional school which assures the theoretical instruction, generally one day per week, and with a company which provides an apprenticeship for the young school-leaver. This alternative training scheme may last two or three years in the industrial world. The scheme has been widespread throughout West Germany and as many as 60 per cent of 16–18 year olds have been included. In Denmark the government attempts to ensure that each school-leaver receives some professional training before entering the job market. In order to meet this objective the authorities attempt to develop existing teaching and training structures, either through practical experience or through schools, to develop a solid base upon which to work. France has been equally creative during the last decade, proposing courses for qualifications and induction, creating new integrated industrial training courses, community work and so forth. Since the early 1970s the various initiatives have included laws on alternative professional training, the emergency plan for the employment of young people of 1986 and various arrangements for young people leaving school without the baccalaureat. In Great Britain the Youth Training Scheme initiated in 1983 offered every young person over 16 the possibility of following two years of professional training comprising at least 20 weeks of instruction plus job experience.

Despite the similar experience of economic and social crisis in the 1980s, no two countries have the same training strategies. Despite this diversity, we can observe the emergence of measures and a certain degree of conformity in terms of the philosophy guiding the general approach to education and training. Although the search for solutions to the problem of youth unemployment will remain essentially a national responsibility, the Community will play an appreciable role from now on thanks to the European Social Fund which will assist the financing of costly measures and will encourage member countries to intensify their efforts to solve these considerable problems.

The need to integrate young people into the workforce, and thus to prevent their economic and social marginalization, has encouraged Ministries of Education to review the functioning of state education systems. Among the measures that have been adopted, the following have proved of particular significance.

First, the trend to prolong the duration of school attendance, which is happening throughout Europe, will serve to reduce the number of premature departures from the education system. This objective has been pursued by raising the school leaving age from 14 to 16. This has happened in France, Denmark and the UK, for example. It has also been a matter of policy to make teaching or professional training obligatory for those pupils not staying on after the normal school leaving age. France, for instance, has committed itself to raising the percentage of young people who sit the baccalaureat to 80.

Second, society in general is seeking to better equip young people to cope with periods of formal education, training or re-training throughout their employment careers, even if they left full-time education without gaining formal qualifications.

Third, there are moves afoot to ensure that school education is not over-academic or over-specialist at too early an age.

# 11

# The Geography of Employment

Christian Vandermotten

In 1987 there were 123,589,000 jobs in the European Community (EC) compared with 112,440,000 in the USA and 59,110,000 in Japan. This represents a rather low employment rate in the EC, 38.2 per cent, compared with 46.1 per cent in the USA and 48.4 per cent in Japan. If one considers only the economically active population aged 15–64 then the relative positions are preserved with the EC at 56.9 per cent, the USA at 69.5 per cent and Japan at 71.0 per cent. These are still rather crude rates which need to be revised in order to take into account the importance of part-time employment and the number of hours actually worked per year, but such correction would without doubt only serve to accentuate the gap between the EC and the other two major economic blocs in the capitalist world.

These figures imply that the gross disparities, not balanced by duration of work, in terms of productivity between the EC, the USA and Japan are less than the differences when considered in terms of productivity per head of the population. Again in 1987, at current prices and controlling for parity of purchasing power, the US gross domestic product (GDP) per capita would be 156 and that for Japan would be 112 if that for the EC were to be indexed at 100. If one further takes into account implicit productivity then the US index is reduced to 129 and that for Japan to 88, lower than that for the EC.

Taking a longer-term perspective may help to clarify the situation (table 11.1). Between 1960 and 1987 employment increased sharply in Japan (+36 per cent) and even more rapidly in the USA (+68 per cent) and, moreover, in both cases it increased more

**Table 11.1** Evolution of employment, population, gross domestic product and implicit productivity in the EC of 12, the USA and Japan

| | *EC of 12 (except former GDR)* | | | | *USA* | | | | *Japan* | | | |
|---|---|---|---|---|---|---|---|---|---|---|---|---|
| *Period* | *Employment* | *Population* | *GDP* | *Implicit productivity* | *Employment* | *Population* | *GDP* | *Implicit productivity* | *Employment* | *Population* | *GDP* | *Implicit productivity* |
| 1960 | 100.0 | 100.0 | 100.0 | 100.0 | 100.0 | 100.0 | 100.0 | 100.0 | 100.0 | 100.0 | 100.0 | 100.0 |
| 1965 | 102.6 | 105.2 | 127.8 | 124.6 | 107.1 | 107.7 | 126.7 | 118.3 | 113.6 | 105.1 | 156.5 | 137.8 |
| 1970 | 104.4 | 108.8 | 159.9 | 153.2 | 118.2 | 113.4 | 145.5 | 123.1 | 116.4 | 119.9 | 291.2 | 250.2 |
| 1975 | 105.0 | 112.1 | 183.6 | 174.9 | 128.1 | 118.2 | 160.8 | 125.5 | 119.9 | 118.6 | 331.2 | 276.2 |
| 1980 | 106.2 | 114.1 | 212.1 | 199.7 | 148.1 | 126.0 | 190.7 | 128.8 | 127.1 | 124.1 | 421.6 | 331.7 |
| 1985 | 104.4 | 115.5 | 228.0 | 218.4 | 159.8 | 132.4 | 219.7 | 137.5 | 133.3 | 128.3 | 510.9 | 383.3 |
| 1987 | 106.5 | 116.2 | 240.0 | 225.4 | 167.7 | 134.9 | 234.2 | 139.7 | 135.7 | 129.8 | 545.6 | 402.1 |
| 1960–75 (%) | +0.3 | +0.8 | +4.1 | +3.8 | +1.7 | +1.1 | +3.2 | +1.5 | +1.2 | +1.1 | +8.3 | +7.0 |
| 1975–87 (%) | +0.1 | +0.3 | +2.3 | +2.1 | +2.3 | +1.1 | +3.2 | +0.9 | +1.0 | +0.8 | +4.2 | +3.2 |
| 1960–87 (%) | +0.2 | +0.6 | +3.3 | +3.1 | +1.9 | +1.1 | +3.2 | +1.2 | +1.1 | +1.0 | +6.5 | +5.3 |

The average growth ratios per annum are computed on the limits of the periods.
*Sources*: EUROSTAT, United Nations and OECD

rapidly than the rate of population growth. By comparison, in the EC the rate of growth of employment was slow (+7 per cent) and less than that of the population as a whole. Combining these data with figures for the growth of the GDP, which is slightly higher in Europe than in the USA but more vigorous than in Japan, it appears that in the long run the USA records only a slow growth in its gross implicit productivity. Once again, this neglects the effects of variations in the aggregate duration of work. This slower growth in the USA is the consequence of having a gross investment ratio systematically lower than in Europe: 18 per cent of the GDP in 1970 against 23 per cent in the EC, and 17 per cent in 1987 compared with 19 per cent in the EC. In the long term this entails a relative setback to technological advance in the USA. Japan, by contrast, is adding to high employment growth and high investment ratios (36 per cent of the GDP was dedicated to the build-up of assets in 1970, and in 1987 this was still 29 per cent), which implies a very rapid increase in implicit productivity.

Such a comparison of these three regional economies enables us to see more clearly the slow down in the rate of productivity increase since the world economic crisis of 1974. It also helps us to appreciate how the EC's economy has fared in comparison with its major competitors.

However, within the EC there is, paradoxically, no correlation between the growth of employment and that of the GDP (table 11.2). In the period 1960–87 the growth of employment was at

**Table 11.2** Evolution of aggregate employment and gross domestic product, 1960–87 (1960 = 100)

| | *Evolution of employment* | *Evolution of GDP* |
|---|---|---|
| Portugal | 148.9 | 340.1 |
| Denmark | 130.3 | 224.8 |
| Netherlands | 128.3 | 237.4 |
| Luxembourg | 116.6 | 225.6 |
| France | 111.6 | 267.9 |
| Greece | 109.0 | 355.7 |
| Spain | 107.0 | 338.0 |
| Belgium | 106.1 | 240.6 |
| UK | 104.3 | 187.3 |
| Italy | 103.3 | 278.1 |
| FRG | 97.4 | 224.3 |
| Eire | 97.4 | 274.7 |

its highest in Portugal – where new economic development was substantial – the Netherlands and Denmark, but in each case substantial use was being made of part-time labour. The other Community countries range from those experiencing a slight decrease in employment, of the order of 3 per cent in Eire and the Federal Republic of Germany, to a growth of 17 per cent in Luxembourg. In the same countries GDP varied from an increase of 87 per cent in the UK to over 350 per cent in Greece.

## Concepts and Methodology

It is clear at the outset that, at both the international and regional scales, levels and patterns of employment are the result of complex interactions between economic and social structures, and that the patterns to be observed, as well as the balance of contributory factors, will appear to vary according to the scale of analysis adopted. The characteristics of employment and unemployment should be considered within actual regional spaces whose economic, social and cultural circumstances are the result of distinctive phases of historical development. The status of the labour market is not simply a mechanistic duplication of the economic situation. It can only be understood in its totality by appreciating the distinctiveness of national and regional ideologies concerning the relations of society and the role of work.

Simple geographical descriptions of employment or unemployment indicators cannot be used directly as guides to the magnitude of regional problems. Other important factors must be taken into consideration, such as the role of the state and its allocation of public subsidies, but also the general relationship between social space and the political superstructure which it supports. It is also important to note the reinforcing element which links current perceptions of space, objective economic conditions and future investment strategies, especially among the most profitable, most powerful and best informed producers. There are also important sociological questions which have a regional or local dimension, particularly the relationship between social classes, parts of classes and institutions together with the ways in which such relations are revealed in the labour market.

In analysing changing employment patterns in Europe it is also important to be aware of the distortions that ensue when one takes instantaneous snapshots of an evolutionary process, especially when the timing of the snapshots is externally controlled and not synchronized among all the constituent parts.

In spite of these various methodological difficulties it is possible to make some progress in the exploration of regional employment structures in the Community. I shall concentrate on the age-group 15–64 and use EUROSTAT data for the mid-1980s in order to explore the geography of employment. However, it will be necessary at the outset to distinguish the following categories of employment: full-time employment; part-time employment; unemployed persons, including persons without employment but actively seeking and immediately available for employment; and unemployed persons not looking for employment, such as rentiers, the handicapped, students, those undergoing military training, women at home and those who have retired early from employment. A substantial proportion of this last category may constitute a potential labour reserve, to be mobilized in response to the ever-changing demands of economic or social circumstances. The last two categories represent the unemployed population while the first two in combination give the employed population in the 15–64 age-group, although a small number aged 65 and over may also be in employment. The employed population plus unemployed persons looking for employment form the labour force and are often referred to as the active population (table 11.3).

If one adds the unemployed persons looking for employment to the 124.4 million Europeans actually in employment in 1986, making up 39 per cent of the aggregate population, then the gross

**Table 11.3** Employment ratio in the countries of the European Community, 1986

| | *Population aged 15–64* | | | *Net employment of population aged 65 years and over (% of the aggregate net employment)* |
|---|---|---|---|---|
| | *Full-time employment (%)* | *Part-time employment (%)* | *Net employment ratio (%)* | |
| Denmark | 58.3 | 17.5 | 67.0 | 1.3 |
| Portugal | 58.8 | 2.7 | 60.2 | 2.0 |
| UK | 50.9 | 13.4 | 57.6 | 0.9 |
| Germany (FRG) | 53.6 | 7.6 | 57.4 | 0.6 |
| Luxembourg | 54.1 | 3.7 | 56.0 | 0.7 |
| France | 51.8 | 6.7 | 55.2 | 0.4 |
| Greece | 50.5 | 2.1 | 51.6 | 2.0 |
| Italy | 49.9 | 2.2 | 51.0 | 0.9 |
| Belgium | 47.8 | 4.8 | 50.2 | 0.3 |
| Eire | 46.5 | 2.2 | 47.6 | 1.9 |
| Netherlands | 41.6 | 10.9[a] | 47.0 | 0.4 |
| Spain | 42.6 | 2.1[a] | 43.7 | 0.8 |

The net employment totals the full-time employment and half the part-time employment.
[a] Estimate.

activity rate is 44 per cent. There were 122.6 million workplaces occupied by people in the 15–64 age-group, thus giving a gross employment rate of 57 per cent and a gross activity rate, including unemployed persons, of 64 per cent. These rates vary by gender: the gross employment rate for men was 71 per cent with 43 per cent for women, while the gross activity rates were, respectively, 79 and 49 per cent.

These gross rates need to be adjusted to allow for the extent of part-time employment. I shall assume that among those in part-time employment and aged 15–64 the average workload is half that of those in full-time employment. This will also be assumed to apply to those aged 65 and over although it is known that in certain Community countries – notably Portugal, Greece and Eire – half-time working may represent a rather low estimate.

These definitions and assumptions allow me to construct a map of the net employment rate among the various regions of the EC

**Figure 11.1** Regional variations in the net employment rate among 15–64 year olds, 1986.

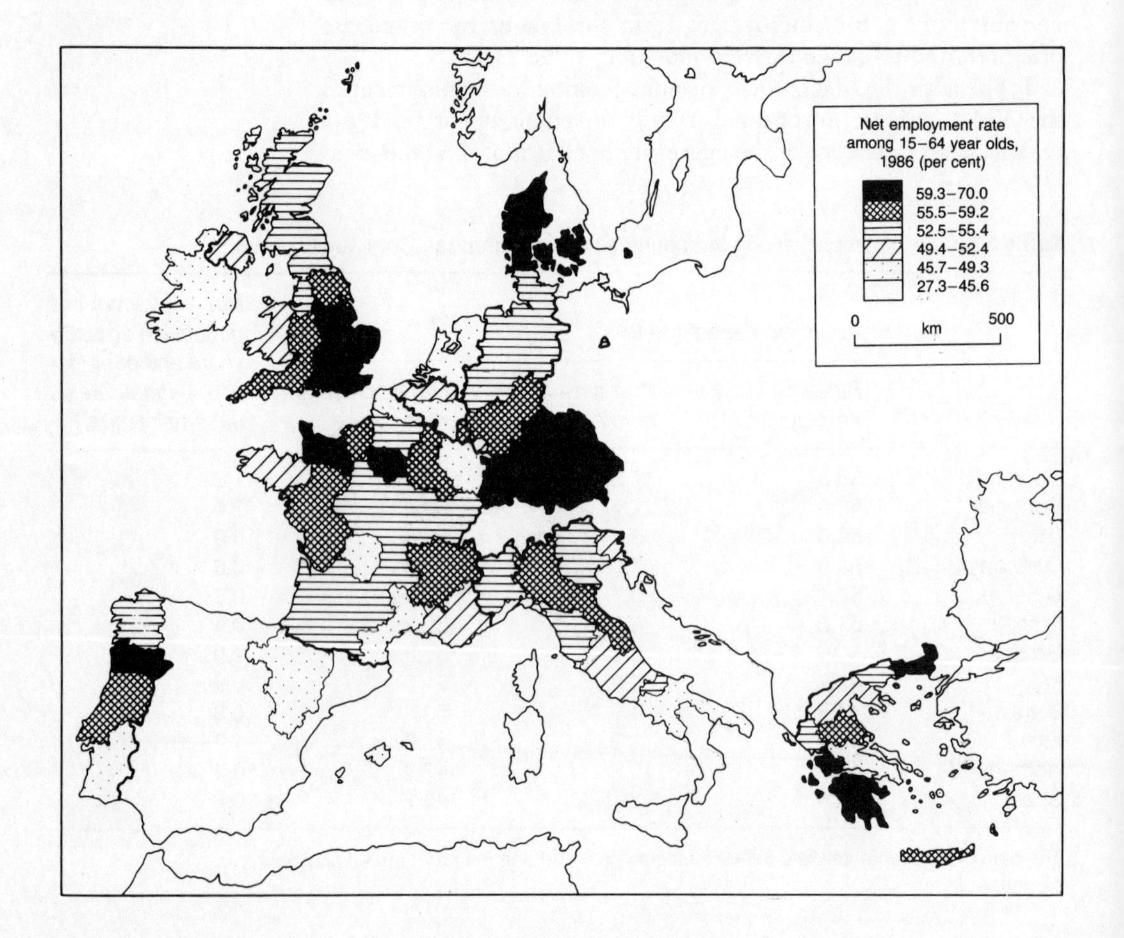

in 1986 (figure 11.1 and table 11.3). There is no simple distinction between core and periphery. Although the net employment rates are high in Denmark, the UK, the Federal Republic of Germany and France, they are low in Belgium and especially the Netherlands; they are also low in Spain and Eire, rather low in Italy and Greece, but rather higher in Portugal. These net rates seem to reflect national practices in the regulation of employment and non-employment, that is, exclusion from the labour force, as well as differences in the balance of full-time to part-time working practices within employment. They are not merely a reflection of levels of economic development.

Within each country the net employment rates are, in general, higher in the most economically dynamic regions and lower in those regions with a traditional industrial base. This contrast is evident if one compares Flanders and Wallonia in Belgium; southern and northern Germany; the south east of England and northern Great Britain; the Paris region and Nord-Pas de Calais; and northern Italy and the Mezzogiorno. But there are certain exceptions to this general point, especially where, as in Galicia and northern Portugal and the peripheral regions of Greece, high employment rates are associated with traditional agrarian economies. Part-time employment is often important, especially in the northern countries of the EC – Denmark, Great Britain and the Netherlands – but it is less important in southern Europe, Belgium and Eire. The Federal Republic of Germany and France occupy intermediate positions (table 11.3).

## Variations in the Practice of Defining Employment and Non-employment

### *Participation In and Exclusion From the Labour Force*

Men aged 25–54 represent the core of the labour force. In all Community countries the activity rate among this group is about 95 per cent, slightly less in Spain (table 11.4). But among women in this age-group there are considerable variations in activity rates: 35 per cent in Spain and 39 per cent in Eire, yet 87 per cent in Denmark.

The significance of female labour may also be shown using the percentage of women in aggregate gross employment (figure 11.2). The regional variations depend as much as those between countries on a number of distinctive and often locally particular factors. In Germany, female employment is most developed in the south, which has a higher economic growth but is also more

**Table 11.4** Gross activity ratios for adult men and women, 1986

| | *Men aged 25–54 (%)* | *Women aged 25–54 (%)* | *Men aged 55–64 (%)* | *Women aged 55–64 (%)* |
|---|---|---|---|---|
| Belgium | 94 | 58 | 41 | 11 |
| Denmark | 94 | 87 | 69 | 46 |
| Germany (FRG) | 93 | 60 | 59 | 25 |
| Greece | 92 | 47 | 65 | 26 |
| Spain | 88 | 35 | 66 | 19 |
| France | 94 | 70 | 44 | 28 |
| Eire | 94 | 39 | 72 | 17 |
| Italy | 96 | 49 | 54 | 15 |
| Luxembourg | 96 | 45 | 40 | 14 |
| Netherlands | – | – | – | – |
| Portugal | 94 | 63 | 65 | 29 |
| UK | 95 | 69 | 68 | 35 |

conservative, while in England the reverse prevails. Female employment is very important in Denmark, but rather low in the Benelux countries. In France it is at its highest in the Ile de France region but especially in those extra-metropolitan localities that have experienced rapid development through industrial decentralization. Female employment is, in general, low in the broad periphery, but there are also important differences within Italy and northern Portugal, with Galicia and parts of Greece proving to be exceptions. The Portuguese example is interesting because the recent growth of the tertiary sector in Lisbon, the development of the textile and garment industries in the north, the mini-fundiary agrarian system and the emigration of adult males may all have been important in particular local circumstances.

Let me emphasize the point once more: even though female employment is actually generated by economic structures, for example via office work in the big metropolitan centres of tertiary employment, through the decentralization of industrial employment to the outer suburbs or even in the workshops of the 'third Italy' or northern Portugal, women only present themselves on the labour market in terms of some nationally determined sociological context. The strength of the Roman Catholic church in Spain and Eire and the social liberation of Scandinavian women each play a part above and beyond purely economic considerations.

The extent of female unemployment must also be considered alongside the degree to which women are engaged in part-time employment. The former tends to be highest in the UK, Denmark,

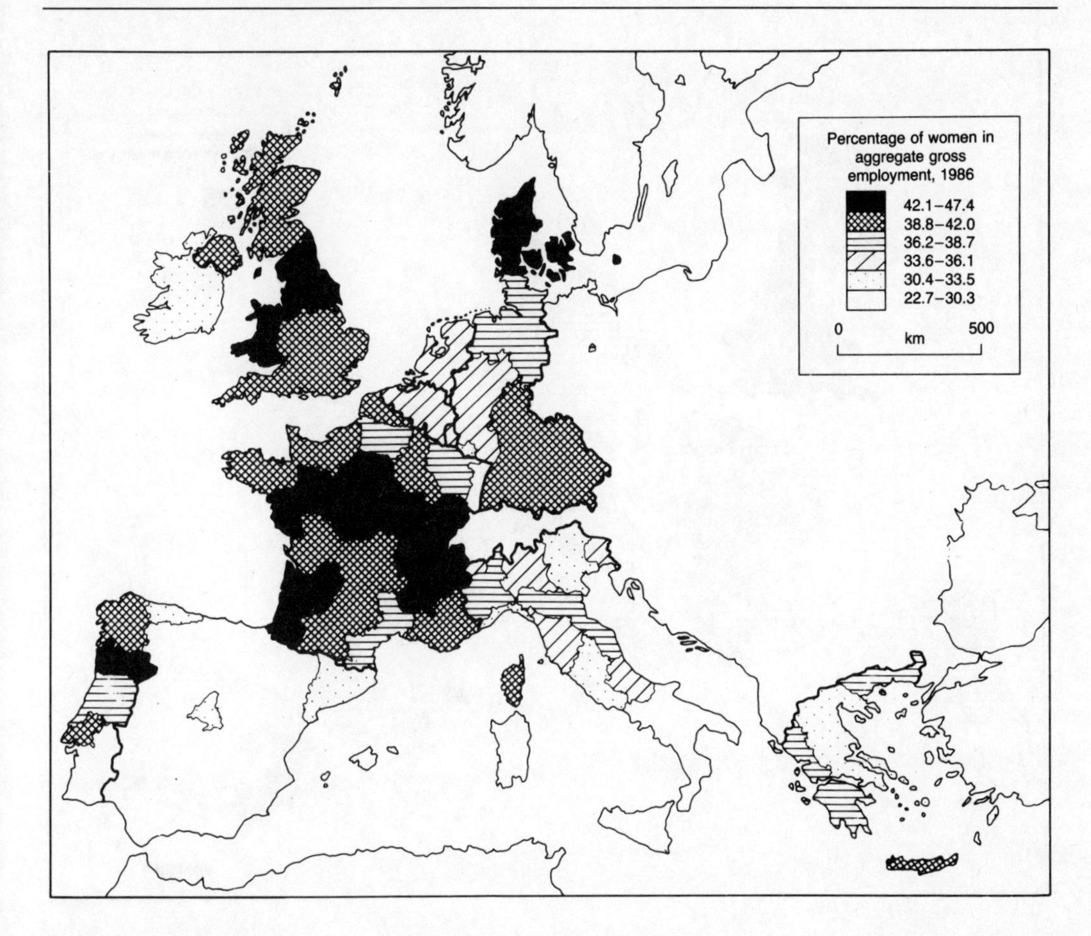


**Figure 11.2** Regional variations in the percentage of women in aggregate gross employment, 1986.

France and Germany. In northern Italy and Belgium, a low or moderate level of part-time employment corresponds to a lower female employment rate in general. On Europe's periphery there are also generally low female activity rates.

Another important aspect of the way in which social attitudes to female employment may vary is reflected by persistence with age. Figure 11.3 shows an index of female employment persistence for 1986. It has been constructed by taking *A*, the ratio of employed women aged 45–54 to employed women aged 25–34, and *B*, the ratio of the total labour force aged 45–54 to that aged 25–34, and finding the ratio of *A* to *B*. A low index suggests a relatively early retirement of women from the labour force while a higher value would imply that women in the post-reproductive years remain in or return to employment. Once again, it is no straightforward matter to interpret the resulting pattern of variation. There is, of course, a positive correlation between female em-

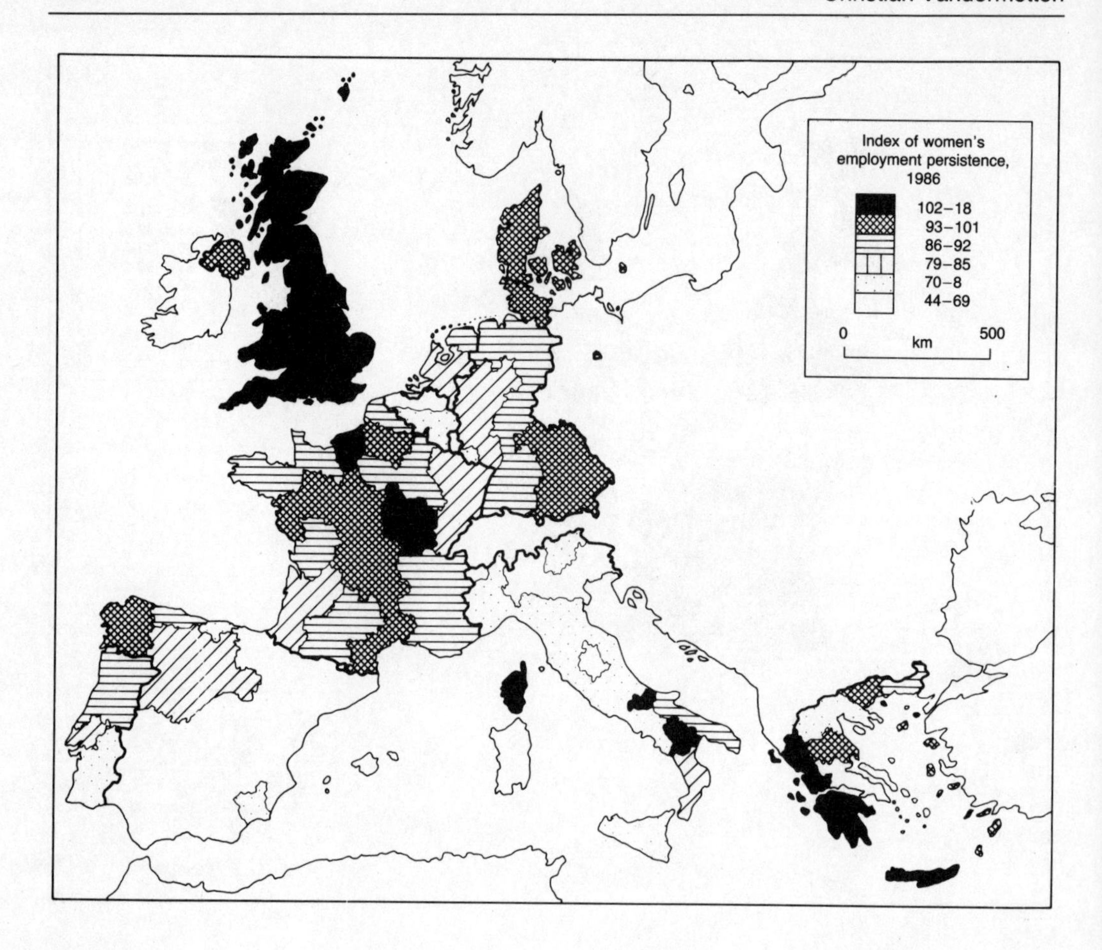


**Figure 11.3** Regional variations in the index of women's employment persistence, 1986.

ployment rates and the persistence index, especially in Denmark, the UK, France and the Federal Republic of Germany. But the cases of Belgium, the Netherlands and northern Italy suggest that where the level of female activity is average and the persistence with age low or only average, then nothing can ensure that young women now included in the labour force will leave it in the near future, while the low rates of female activity at an older age may reflect the fact that many women have not in the past been present on the employment market. There is some association between those countries with low female employment rates in the 1960s which have subsequently experienced some of the highest rates of growth. In southern Italy, north western Iberia and parts of Greece the picture may also have been confused by high levels of male emigration.

In recent years the activity rate for those aged 55–64 has declined dramatically. This rate, which is particularly low in

France and Belgium, is influenced by the prevailing retirement age and the variable significance of non-salaried employment. It tends to be higher in Europe's peripheral regions and lower in the traditional industrial regions whose economies have undergone radical restructuring often accompanied by substantial unemployment and early retirement.

At the other end of the age range, the activity rate among young people aged 15–24 is relatively high in Denmark, the UK and the Federal Republic of Germany and is low in Spain, Greece and Italy. Low activity rates in this age-group will be affected by regional economic difficulties and the general state of the job market, but they will also be influenced by the tendency to prolong formal education which often facilitates access to the tertiary sector as well as emigration from the home region when there are insufficient local opportunities for well-qualified young people.

### *Unemployment*

At first glance, the highest unemployment rates appear to lie in the periphery of the EC (figure 11.4), while the lowest rates prevail in the new economic heart of Europe, southern Germany. But within the old industrial heartland of Europe there are also regions or sub-regions with particularly high unemployment, ones that are equivalent to parts of the periphery. North Rhine-Westphalia, Wallonia, Nord-Pas de Calais and northern Britain are particularly good examples of regions whose economies were built in the nineteenth century on coal, iron and steel, heavy industry and mechanical engineering. Within the old core the regions with the lowest unemployment rates tend to be the great metropolitan centres in which tertiary sector employment has remained buoyant. The metropolitan regions of the Federal Republic of Germany, France, the UK and Italy each had substantially lower unemployment rates than the national average, but in Greece, Spain and Portugal the equivalent regions had equal or higher employment rates.

Outside the Community, unemployment rates are high in the east but low in the north to the extent that the simple notion of core and periphery does not hold particularly well. Even in the EC periphery there are regions which have attracted new forms of investment and employment, although the traditional solution has come via inter-regional migration or international emigration.

It also appears that large areas of the Community combine a low unemployment rate with high female participation in the workforce while, particularly in the peripheral regions, high un-

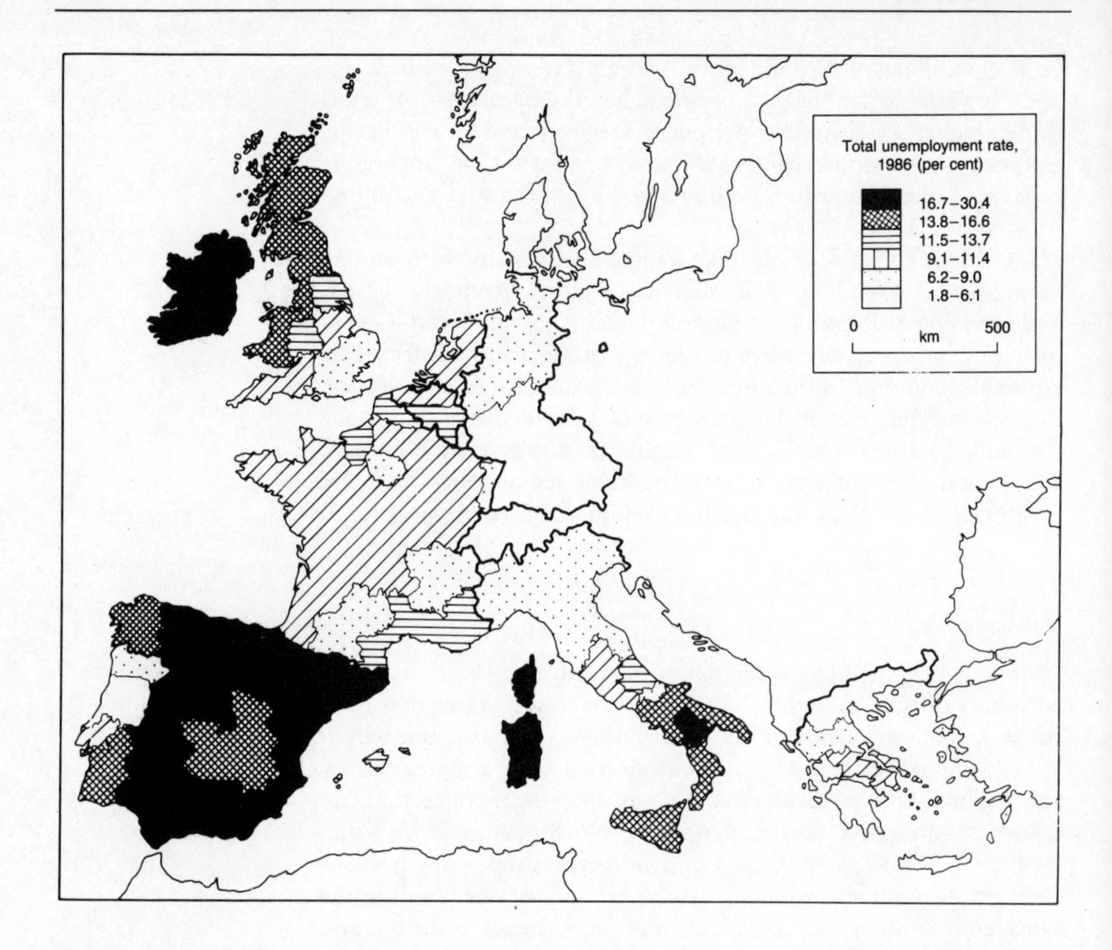


**Figure 11.4** Regional variations in total unemployment rates, 1986.

employment may be combined with a low female participation rate.

Chapter 12 takes up the theme of regional variations in unemployment rates and considers them in more detail.

### A Typology of Employment/Non-employment

Thus far the dividing lines between employment, unemployment and non-activity have not provided the simple means to distinguish between core and periphery as one would have expected. A rather more sophisticated typology is required, one that recognizes the importance of social practices developed within national boundaries. I shall propose here a five-element classification.

1 Spain and Eire show very low employment rates, especially among women but also among men. Non-employment takes

the form of non-activity among women and high unemployment among men.
2 The Benelux countries, Italy and Greece have employment rates close to the EC average, but female non-employment is high.
3 In France, the Federal Republic of Germany and Portugal the male employment rate is also close to the EC average, but female non-employment is lower except, that is, in the old industrial regions (North Rhine-Westphalia, Nord-Pas de Calais, Lotharingy). The higher than EC average female employment rates in France and the Federal Republic of Germany can be linked with the popular recourse to part-time employment, but this form of employment is less frequent in Portugal.
4 In the UK the male employment rate lies near the EC average, but unemployment is high, especially outside the prosperous south east. The employment rate among women is slightly higher than the average and unemployment is not excessive, but part-time employment among women is highly significant.
5 Denmark provides an example of a country with a very high employment rate among men and an exceptionally high employment rate among women. Unemployment is low, but part-time employment is very significant.

## Employment Structures

Although it was not a straightforward matter to distinguish core from periphery by only using the distribution of employment and non-employment, the use of employment sectors may prove more appropriate.

Beyond the metropolitan areas of Portugal, Spain and Greece agricultural employment is still especially important with from 15 to 25 per cent of the active population so engaged. In southern Italy, southern Spain, western France and Eire this agricultural sector is supported by the public tertiary sector and/or tourism. Within Europe's economic core area there are differences between those regions that still rely heavily on industrial employment, up to 40 per cent, and those regions dominated by the tertiary employment with 70–85 per cent in that sector. The most recently developed, more diversified and in general more dynamic regions of the economic core (southern Germany, Piedmont, Catalonia, for example) also tend to have a smaller residual agricultural sector compared with the older industrial regions (English Midlands, North Rhine-Westphalia, Lombardy). An extreme case of structural crisis in an old industrial region is that of Wallonia which is undergoing what could be called 'tertiarization by default' with the disappearance of its industrial and mining potential. The

French economy appears to have a larger tertiary sector than the German economy even in its old industrial centres of Lotharingy, Nord-Pas de Calais and Lyon, but one must also take into account the tendency for German firms to internalize their service requirements. In the Paris region, where industrial dispersal has been prominent, and in the 'third Italy' the agricultural sector still retains about 10 per cent of actively employed persons. The regions of northern Portugal, Navarre and to a lesser extent Aragon and the Cantabric coast represent a special case in which the residue of old established industries co-exist with traditional agrarian structures. All the great metropolitan regions – Paris, Brussels, Randstad Holland, Copenhagen, Madrid, Rome and London – are parts of Europe with the most highly developed tertiary sectors. The south west of England and Provence–Côte d'Azur also belong to these tertiary, tourist and residential areas.

On the international scale these sectoral structures display a tendency to homogenize over the long run. Figure 11.5 uses an employment triangle to trace the changing balance between the agricultural, industrial and tertiary sectors between 1930, 1950, 1960 and 1987. The first phase of change stopped in the early 1960s, except in the UK and Belgium where it began a good

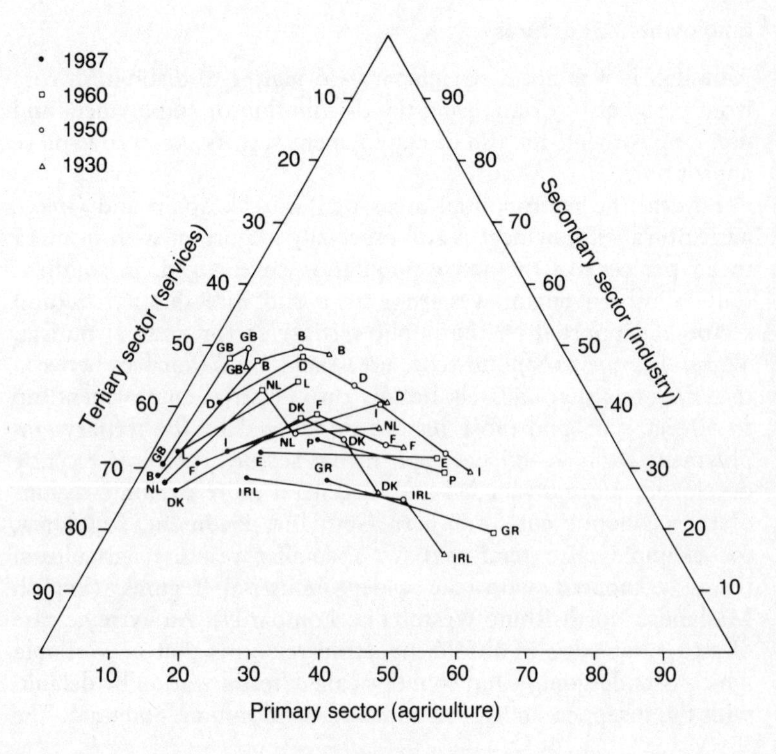


**Figure 11.5** National variations in the changing balance between the primary, secondary and tertiary sectors of employment.

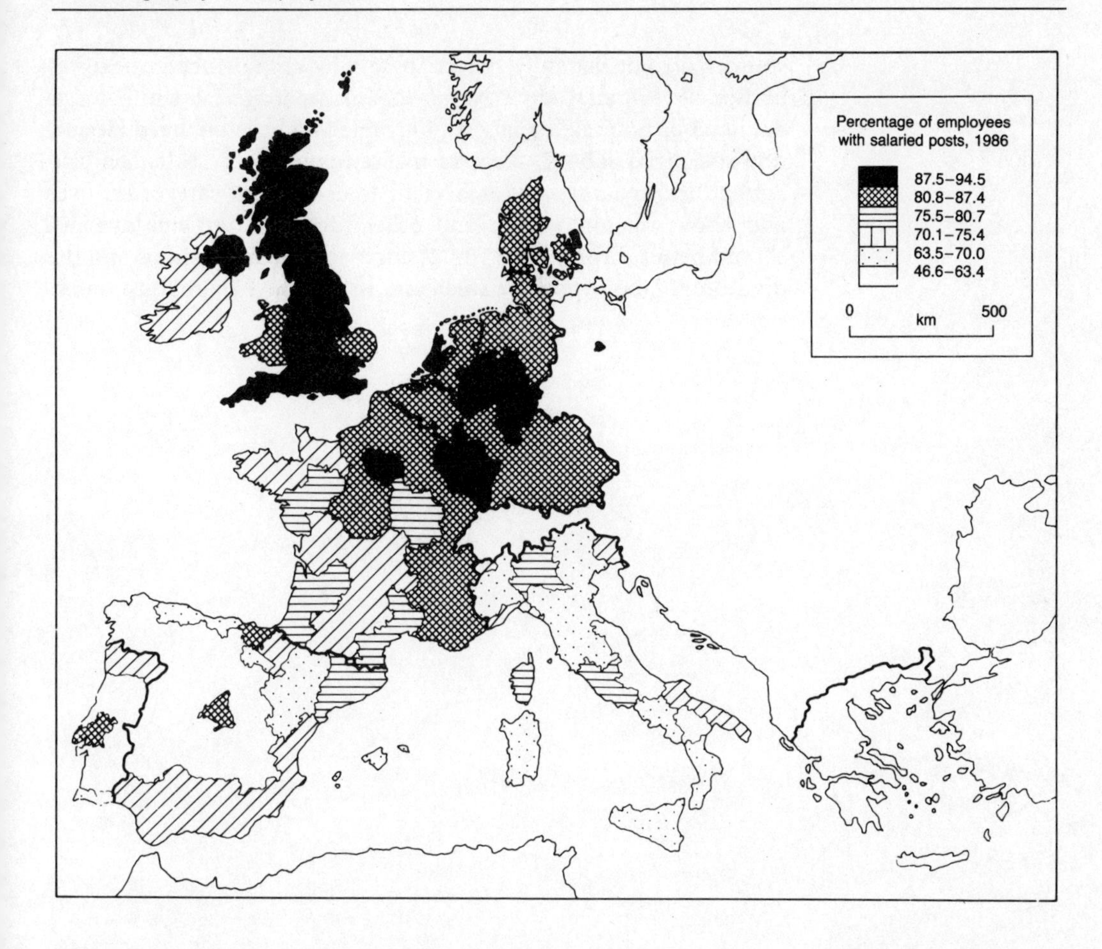


**Figure 11.6** Regional variations in the percentage of employees with salaried posts, 1986.

deal earlier, but in Eire, Greece, Portugal and even in Spain de-industrialization has barely begun.

This analysis in terms of only three employment sectors is becoming increasingly unrealistic as the new technologies lead to complex inter-penetrations between the industrial and tertiary sectors. Within firms there is a growing tendency to concentrate production in the low-cost labour pools – whether these are in the Community, in eastern Europe or the Third World – to focus research and development facilities in the outskirts of the great international metropolitan centres and to locate headquarter offices in the centres themselves.

Some sense of these emerging differences may be gained by considering variations in the percentage of salaried employees. Levels are particularly high in the UK, and east of a line from Le Havre to Marseille, but in Iberia, Italy, Greece and Eire rates are still rather low (figure 11.6). Unfortunately, European statistical sources do not yet fully reflect these new employment categories in the new spatial division of labour: managerial staff, stable qualified or unqualified labour, unstable and little qualified labour, non-socialized labour such as the self-employed, craftsmen and agricultural tenants. Any analysis based on these categories, over and above unemployment and other forms of non-employment, should better display the form of core–periphery relations and the division of labour both between and within the Community states.

# 12

# Unemployment: Regional Variations in Age- and Sex-specific Rates

Aurora Garcia Ballesteros

Labour imbalances are mainly due to contradictions in the economic system which are in turn closely linked with demographic variations in the field of employment. Largely as a consequence of the economic crisis, geographers have become more interested in analysing the conditions as well as the regional and local divisions of labour (Gambier, 1980; Vandermotten and Grimmeau, 1983, 1985). They also want to know whether unemployment rates present spatial differences during periods of economic recession (Fischer and Niskamp, 1987).

Of the various theories and definitions of labour market segmentation (Morrison, 1990), the most appropriate for our purposes comes from the Commission of the European Community (1981). Their definition focuses on the procedure by which the labour market is divided into subsets with different features and working rules. It implies the marginalization of certain groups of hired labour. Among the subsets, gender and age relations have their own particular importance. People looking for jobs will always present certain personal characteristics which may help or hinder their inclusion in the labour force. Some of the most important characteristics are age (excessive youth or age may prove detrimental), sex (women are the labour army in reserve), ethnic origin, training and so forth.

It is difficult to compare regional variations in unemployment rates by age and sex. The most valuable economic units do not always coincide with the administrative units adopted by the European Community (EC) and, moreover, the European countries have different definitions of unemployment (see also chapter 11).

However, the statistics provided by EUROSTAT make possible some comparative analysis on a regional scale.

### Unemployment in the European Community

High unemployment rates appeared in the EC during the first economic crisis of the 1970s. Until then unemployment in the 12 countries had been less than 3 per cent. The most dynamic economies absorbed immigrants from those countries with high demographic surplus and a largely agrarian economic base, such as Spain and Italy. In 1973, the Community's average rate of unemployment was 2.8 per cent, although there were high rates in Eire with 5.7 per cent and Italy with 6.4 per cent. The Spanish unemployment rate became the highest in the Community in 1985 with 21.9 per cent, when the average was 10.9 per cent.

In the 1970s, the second economic crisis of the last few decades removed a large number of jobs, but in the 1980s the gross national product grew, jobs were created at about 1 per cent a year and unemployment fell in consequence. In 1990, the unemployment rate was below 9 per cent (Uner, 1991). Between 1983 and 1990, about 8 million jobs were created compared with 3 million that were lost after the second oil crisis of 1979–80. Apart from Eire, all the Community states have either recovered or exceeded the level of 1980.

Compared with this generally favourable trend, some countries have experienced only a slow decrease in unemployment. The rate even increased in the first-half of the 1980s. In 1973 the unemployment rate was 2.8 per cent with 6.4 per cent in 1980, 10.9 per cent in 1985 and 8.7 per cent in 1990. This means that unemployment is still one of the EC's main problems. It is made worse by a particularly high unemployment rate among young adults, about 20 per cent for 25 year olds and the existence of some 5 million long-term unemployed.

The relative economic improvement meant that other population groups began to enter the labour market. This new element of participation, together with the pressure exerted by recent larger generations of young people, helps to account for the continuing high unemployment rates. We must also remember the high female activity rate.

There are, however, persistent differences in unemployment rates between the 12 countries, differences that show little sign of decreasing. In 1980, Luxembourg had the lowest unemployment rate with just 0.7 per cent, while Spain had the highest, 11.6 per cent. In 1990, Luxembourg had 1.7 per cent and Spain had 16.5 per cent; the difference had increased from 10.3 to 14.8 percentage points. The reasons may be found in the economic and demographic conditions, in the participation of women in the labour market and in several other maladjustments in the structure of the labour market (Noin, 1988).

Table 12.1 shows several interesting differences in age- and sex-specific unemployment rates, although the age data only refer to those less than 25. The higher unemployment rates among women reflect market segmentation, especially in Spain where women entered the labour market much later than in other Community countries (Garcia Ballesteros et al., 1985). The UK provides an exception – there female unemployment rates used to equal those of the Community (Salt, 1985).

In general, the unemployment rate among young people is much higher than the country's average; only the Federal Republic of Germany had a slightly lower rate in the late 1980s. Some countries, most notably Italy, Spain and Greece, have very high rates which are probably influenced by their experience of higher demographic pressure (Garcia Ballesteros and Crespo Valero, 1988; Garcia Ballesteros, 1991). Once again, women are the group most discriminated against, especially in Spain and Italy where their efforts to enter the labour market are being partially frustrated.

**Table 12.1** Age- and sex-specific unemployment rates in the European Community

| | *1989* | | *Younger than 25 years* | | | *Total* | | |
|---|---|---|---|---|---|---|---|---|
| | *Men* | *Women* | *Total* | *Men* | *Women* | *1985* | *1989* | *1990* |
| Belgium | 6.4 | 14.8 | 17.8 | 12.7 | 23.1 | 11.7 | 9.6 | 8.8 |
| Denmark | 6.0 | 8.2 | 10.0 | 9.0 | 11.0 | 7.6 | 7.0 | 7.6 |
| Germany (FRG) | 4.6 | 7.4 | 5.3 | 4.8 | 6.0 | 7.3 | 5.7 | 5.4 |
| Greece | 5.8[a] | 13.2 | 27.4 | – | – | 8.7 | 8.5 | 8.5 |
| Spain | 12.8 | 25.1 | 33.6 | 26.7 | 41.9 | 21.9 | 17.0 | 16.5 |
| France | 7.2 | 12.8 | 20.4 | 16.4 | 24.4 | 10.3 | 9.6 | 9.1 |
| Eire | 16.4 | 18.9 | 23.9 | 25.3 | 22.2 | 18.4 | 17.2 | 16.2 |
| Italy | 7.4 | 17.1 | 31.9 | 26.2 | 38.5 | 9.4 | 11.0 | 10.6 |
| Luxembourg | 1.4 | 2.9 | 4.1 | 4.0 | 4.1 | 3.0 | 1.9 | 1.7 |
| Netherlands | 6.8 | 13.2 | 13.7 | 12.9 | 14.5 | 10.4 | 9.3 | 9.6 |
| Portugal | 3.4 | 7.2 | 11.1 | 7.9 | 15.0 | 8.5 | 5.0 | 5.2 |
| UK | 7.0 | 6.0 | 8.9 | 10.2 | 7.3 | 11.5 | 6.5 | 6.5 |
| EC | 7.1 | 11.8 | 17.2 | 14.8 | 19.9 | 10.9 | 9.0 | 8.7 |

[a] Data are for 1988.
*Sources*: EUROSTAT, various years; author's own estimates

### Regional Variations in Unemployment Rates

Although there are standard data for the Community's regions, the surface area, demographic importance, population distribution and mobility vary from country to country among the 12 member states. Further, the geographical divisions that have been defined for administrative and statistical purposes do not coincide with labour market areas; indeed the statistical regions often display significant internal divisions.

The following conclusions may be drawn from an analysis of regional variations in unemployment rates in the Community. First, there is considerable variation in the rate. The ratio of minimum to maximum among regions tends toward zero at 0.06 in 1987, which reflects a considerable degree of diversity within the Community (Fischer and Niskamp, 1987). Second, there is a good deal of heterogeneity in the distribution of unemployment rates with similar values applying to regions with very different patterns of development. After the first oil crisis of 1973–4 there was a general increase in unemployment rates, but the worst effects were felt in those regions that experienced rapid economic restructuring and where, as in the Spanish regions, out-migration became more difficult. The second oil crisis of 1979–80 even affected those regions that had until then remained prosperous. Their unemployment rates increased to previously unexperienced levels. That is why similar rates may be experienced in regions with rather different employment dynamics and problems.

The lowest regional unemployment rates, below 6.5 per cent, are particularly concentrated geographically. Low values are to be found in the Federal Republic of Germany; parts of Denmark; Luxembourg; parts of Greece, especially the islands; Italy, especially Lombardy and Aosta; and some areas of Portugal. The higher than Community-average levels of unemployment are to be found on the periphery, especially Spain, southern Italy, Eire, northern Britain and Northern Ireland, but there are certain regions of relatively high unemployment in each country. In the Federal Republic of Germany the higher rates are to be found in the port cities of Bremen and Hamburg. In Belgium, the Walloon region contrasts with Flanders. In France, the highest rates are to be found in Nord-Pas de Calais and the south, while in the Netherlands it is the north that has the relatively higher rates. Again, the simple rates may be similar but the underlying circumstances will vary.

Because of the influence of the two oil crises, the highest unemployment rates appear in those regions where the primary sector prevails (see figure 11.5). In these regions the upward trend of job losses has persisted although there have been short-term fluctuations that have coincided with temporary up-swings in the national economies. Traditionally, these regions had highly mobile populations – emigration acted as a safety valve to relieve the intense pressure placed on the labour market by successive generations of young workers. Economic crisis in regions that had in the past normally received those emigrants, and the consequent disruption to the migration chain, tends to push unemployment in the sending regions even higher. Besides, the new local industries were unable to absorb the surplus agricultural population and the demographic pressure from the general population pressure that came from high rates of natural population growth. The service sector of the economy only developed in a localized fashion and then was often associated with the seasonal employment generated by tourism. Andalucia, Campagna and Sicily provide good examples of this sequence of events and its creation of high unemployment.

High unemployment rates are also to be found in industrial and service areas; in many cases these are urban regions that formerly received substantial numbers of migrants. The recent economic crises have particularly affected those regional economies that were insufficiently diversified and the subsequent restructuring has meant substantial job loss. Between 1973 and 1981 manufacturing industry in the UK lost 1.2 million jobs and in the 1979–81 crisis gross investment decreased by a further third. Companies were therefore obliged to cut jobs even further (Martin, 1982; Salt, 1985; Wabe, 1986). High unemployment rates in the English Midlands, Yorkshire, Humberside, Wallonia, Catalonia and the Basque region – which has also been affected by socio-political troubles – may be due to this combination of factors, especially if the service sector, a key contributor to new job creation, has not grown at the same pace. Besides, unemployment may have decreased in some areas where it has been persistently high because people have been discouraged from even entering the labour market. Thus, the activity rate has fallen and the unemployment rate has increased. This is the case in southern Italy, Spain and some areas of Belgium, the Netherlands and Ireland, north and south.

The lowest unemployment rates coincide with high activity rates and are to be found in the most diversified and dynamic industrial areas of Europe, where there has also been an important increase in productive and user services. In some urban areas, expansion in the service sector is the only reason for the steady growth of employment. Examples are to be found in most parts of southern Germany, together with the Ile de France, the Po Valley, Luxembourg, northern Belgium and the south east of England.

**Table 12.2** Inter-regional minimum/maximum ratios for unemployment rates, 1987

| | *Total* | *Men* | *Women* | *Youths* | *Adults* |
|---|---|---|---|---|---|
| Belgium | 0.45 | 0.34 | 0.45 | 0.34 | 0.48 |
| Denmark | 0.65 | 0.78 | 0.55 | 0.72 | 0.63 |
| Germany (FRG) | 0.26 | 0.21 | 0.30 | 0.20 | 0.27 |
| Greece | 0.29 | 0.29 | 0.26 | 0.30 | 0.26 |
| Spain | 0.40 | 0.28 | 0.33 | 0.46 | 0.34 |
| France | 0.48 | 0.44 | 0.50 | 0.42 | 0.49 |
| Italy | 0.27 | 0.23 | 0.29 | 0.25 | 0.24 |
| Netherlands | 0.46 | 0.43 | 0.51 | 0.44 | 0.47 |
| Portugal | 0.31 | 0.26 | 0.26 | 0.26 | 0.23 |
| UK | 0.42 | 0.38 | 0.48 | 0.43 | 0.43 |
| EC countries | 0.08 | 0.06 | 0.10 | 0.05 | 0.06 |

*Sources*: EUROSTAT; author's own estimates

There are also low unemployment rates in certain traditional agricultural regions which have had a long history of emigration. The Massif Central, Abruzzi-Molisye, parts of Portugal and Greece would all fall into this category. Here low unemployment rates go hand-in-hand with demographic decline (see chapters 14 and 15).

Regional contrasts are also revealed in table 12.2. The minimum/maximum ratio among regions varies from Denmark with 0.65 to the Federal Republic of Germany with 0.26. There are only slight variations in the former, but in the latter Bremen and Hamburg contrast sharply with areas of southern Germany. In general, intra-national contrasts are not as great as those between countries.

## Regional Variations in Unemployment among the Young

Unfortunately, it is not a simple matter to define youth unemployment among the 12 Community countries. There are differences in minimum legal working age, compulsory school levels, training levels and company qualifications as well as differences in the numbers of new workers. Figure 12.1 gives a general impression of the rate of youth unemployment for those aged under 25 for the year 1987.

The youth unemployment rate varies widely in the Community as the minimum/maximum ratio in table 12.2 reveals. At the Community level, more than one-fifth of the under-25s do not have a job. Only in the Federal Republic of Germany, Denmark

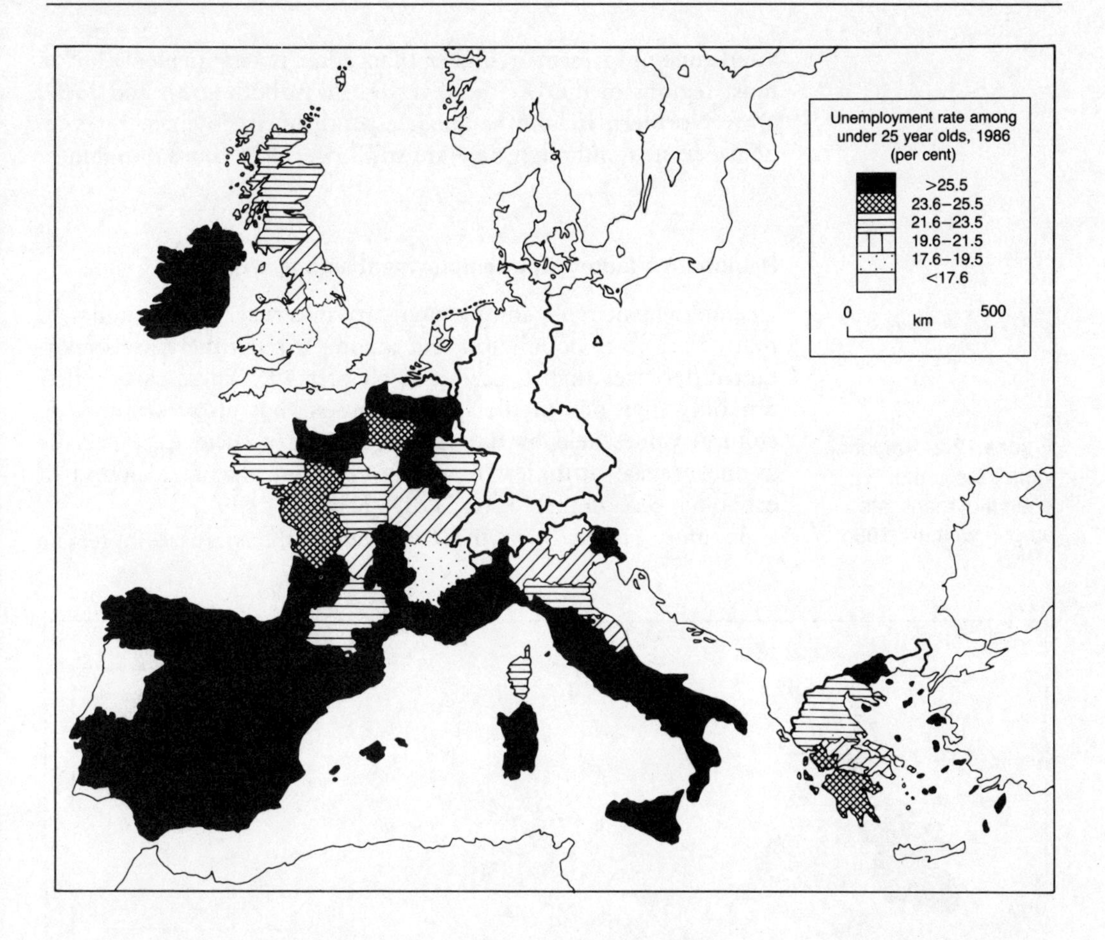


**Figure 12.1** Regional variations in the youth unemployment rate (under 25 year olds), 1986.

and, but to a lesser extent, the Netherlands and the UK (although Northern Ireland has high rates) are there lower rates. All the other countries have regions with serious youth unemployment. In Belgium, in the Walloon area the rate is over 30 per cent and contrast with the north of Belgium is especially sharp: there the rate is below the Community average. France and Portugal show equally striking regional contrasts.

The problem is getting worse in three countries: Eire, where there are also high birth rates; Spain, where all regions have high youth unemployment rates, but especially Andalucia with 53.8 per cent and the Basque country with 58.9 per cent (Garcia Ballesteros et al., 1985); and Italy, which has the region with the highest rate, Campagna with 63.2 per cent, although in general rates are not as high as those in Spain. In many of the worst affected areas, emigration has been curtailed, but demographic pressure persists.

The UK is a special case. In other countries of the Community youth unemployment is higher than adult unemployment, but in most regions of the UK the reverse held in both 1986 and 1987. Only Northern Ireland has higher youth unemployment rates, at 26.5 per cent, although they are still lower than those of Spain or Italy.

#### Regional Variations in Unemployment among Women

Unemployment rates among men vary in a way that is similar to that of the total population, but among women there are important differences (figure 12.2) (see chapter 13). These rates reflect not only divisions in the labour market, but also variations in cultural values held by the various European societies, especially as they reveal attitudes to women working during a period of economic recession (Fincher, 1989; Beneria, 1990).

In most countries of the Community the increase in female

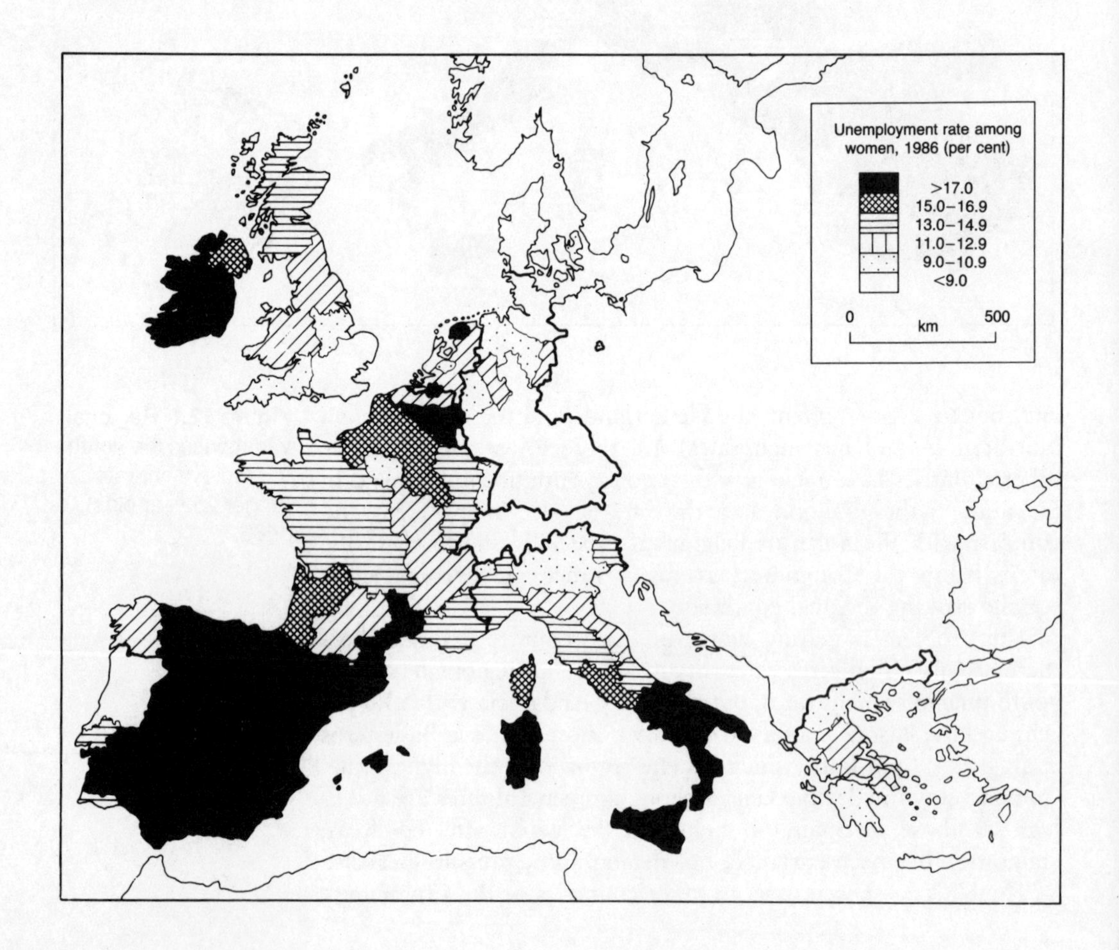


**Figure 12.2** Regional variations in the unemployment rate among women, 1986.

activity rates (over 50 per cent in some areas since 1980) has coincided with the recent economic recession. Some governments, that of Eire for example, have initiated policies that promote men's work rather than women's (Pyle, 1986). The employment of women has increased in the industrial sector which has been particularly hard hit by the recession and in the service sector which is weathering the crisis better.

Regional variations are also important, although they are not as clear as in other population sub-groups. The Community's unemployment rate among women is 13 per cent, 4 percentage points higher than that among men. The higher values occur in those Spanish, Belgium, French and Italian regions with high industrial unemployment rates or in those areas with traditional agrarian patterns where female agricultural labour is not always recognized formally. In such places women have been obliged to take the place of men who have emigrated. However, in most regions female activity rates are low and the current economic crisis has discouraged women from entering the labour market, thus keeping unemployment rates constant.

Once again, the UK is a special case. In 1986 and 1987 unemployment rates for women in almost all regions were lower than among men, which are some of the highest in the Community (see also chapter 11).

### Concluding Remarks

Unemployment has become a fact of life that affects almost one-fifth of the EC's population. Its significance can hardly be overstated in terms of its economic, social, political and even directly demographic impact. In regions with chronic high unemployment a new informal, 'black' economy, with its own social fabric and supporting values, has emerged.

Supply and demand in the labour market are out of phase. The reasons for this are partly economic (maladjustments in the productive system), partly demographic (mobility, migration and irregularities in the age–sex pyramid) and partly social (discrepancies between training qualifications and company training requirements, women's work etc.).

Labour market segmentation particularly affects two groups: women and young people. In most European countries these groups experience increasing unemployment, even in those regions experiencing relatively rapid growth. These two groups must be the target populations for Community policies aimed at reducing regional variations in levels of unemployment.

Finally, it is important to consider the local level in analyses of unemployment simply because aggregate statistics for some of the larger regions may be just as misleading as national rates. For example: in Spain, Aragon has the lowest unemployment rate but within the region there is a stark contrast between Zaragoza, an urban centre that has experienced rapid development in recent years, and Huesca and Teruel which are virtually depopulated, the victims of chronic emigration. In Catalonia, just within the boundary of Barcelona, textile manufacturing districts which are suffering from severe economic crisis co-exist alongside advanced technology parks.

# 13

# The Role of Women in the Post-industrial Economy

Giovanna Brunetta

Of the profound technological, economic and social transformations which have influenced the world of work, one particularly significant element has involved the change in the status of women: their progressive entry into the labour force has been one of the most important aspects. In 1950, the typical picture of men at work and wives at home accounted for 70 per cent of the labour force, while today this is only true for 15 per cent. More than 50 per cent of all wives now work (Bell, 1986, p. 58). There have been considerable repercussions on the economic and social planes, particularly in the sphere of the family and reproduction.

## The Increase in Female Labour

In the 12 European Community (EC) countries, the participation of women in the world of work has progressively increased at speeds that have begun to accelerate since the 1970s. The activity rate increased from 24 per cent in 1950 to 25.8 per cent in 1970 and rose to 33.3 per cent in 1987 (table 13.1). The German Democratic Republic followed the same trend but at a rather different pace since even in 1950, in keeping with the dominant social ideology (Schaffer, 1981, pp. 56–8), the female activity rate was quite high at 40.3 per cent. Increased participation of women in the labour force corresponded with a reduction in the male labour force (58.1 per cent in 1970, 55.2 per cent in 1987). Although the two components have been converging, the activity rate among females is still much lower than that among males.

However, not only the supply but also the demand for women's

**Table 13.1** Activity rates among women, 1950–87 (per cent)

| | *1950* | *1960* | *1970* | *1980* | *1987* |
|---|---|---|---|---|---|
| Belgium | 19.00 | 20.10 | 22.35 | 26.55 | 34.10 |
| Denmark | 32.30 | 28.75 | 34.55 | 45.75 | 51.00 |
| Germany (FRG) | 31.90 | 32.60 | 30.60 | 33.60 | 34.70 |
| Greece | 17.95 | 18.90 | 19.45 | 19.50 | 27.50 |
| Spain | 11.55 | 13.40 | 13.40 | 15.95 | 24.20 |
| France | 28.30 | 28.10 | 30.10 | 33.80 | 35.90 |
| Eire | 22.95 | 20.10 | 20.10 | 20.90 | 22.60 |
| Italy | 21.00 | 21.15 | 21.90 | 23.65 | 29.40 |
| Luxembourg | 26.40 | 22.45 | 20.30 | 26.55 | 30.00 |
| Netherlands | 18.65 | 15.75 | 19.05 | 23.75 | 28.20 |
| Portugal | 17.00 | 13.25 | 18.50 | 29.85 | 38.70 |
| UK | 25.35 | 28.70 | 31.85 | 36.25 | 39.40 |
| EC of 12 | 24.00 | 24.74 | 25.84 | 29.13 | 33.30 |
| Germany (GDR) | 40.30 | 43.60 | 42.90 | 47.25 | – |

*Sources*: ILO, 1986; ILO, 1988 for 1987 data

work has increased. The percentage of employed women out of the total employed population increased from 32.5 per cent in 1970 to 37.9 per cent in 1987, although female employment within the Community is lower than that in other industrialized countries such as the USA (44.1 per cent) and Japan (39.9 per cent). But, the mean value for the EC conceals distinct inequalities both between countries and among regions (see figure 11.2). The extraordinary variety which is so characteristic of women's work does not appear in men's. This may mean that the economic role of women depends much more on the environment than that of men, in the sense that male employment filters out and regulates female employment to a greater or lesser extent. However, it is not easy to understand the basic reasons for this differentiation since economic motivations, linked to the actual availability of jobs, are mixed with factors involving customs, tradition and culture. The economic duality which distinguishes the Mediterranean from west-central Europe is a cultural fact which has its roots in the historical forces which contributed towards the maintenance of marked differences between the two areas, partly in terms of the role women play in the job market. The employment rate among women is close to 40 per cent in France, the Federal Republic of Germany and the UK, but it falls to around 30 per cent in Greece, Italy and Spain. The relatively high rate of female employment in certain Portuguese regions may only be understood when it is related to the high proportion of women, more than 50 per cent, working in agriculture. In Italy, there are contrasts between the high levels of female employment in the north and the far lower level in the south. In southern Italy, custom overlaps with economic backwardness. But even in West Germany there are contradictions between the economically dynamic regions of the south, especially Bavaria and Baden-Wurtenberg, and the older industrial regions which have experienced economic crisis, the Ruhr and the Saar.

Women have not only begun to put pressure on the job market, but they have also changed the cycle of their working lives. Trends in activity rates have not only risen considerably (except at the extremes, owing to more generalized and prolonged education and early retirement schemes), but have progressively been losing their characteristic 'hump' which shows the presence of women at the margins of the job market, more in keeping with their traditional family and household commitments than regular, continuous working careers. The trend of women's participation levels has become more and more similar to that of men; women, like men, increasingly enter the job market as young people and leave it on retiring. The reasons for this transformation, this 'masculinization', must be sought in a number of cultural and social events – such as the emancipation of women, the transformation of the family and the fall in the birth rate – which have changed women's attitudes towards work. Increased demand in the services sector, in particular for those professions which by their very nature (less physical effort, flexible daily routine) allow women both to work and to maintain their role in the home, has also been important.

However, the growth in adult activity rates in Community countries is rather diverse. In countries like Eire, Luxembourg and Spain, women's economic activity begins at the end of compulsory schooling, peaks among the 20–4 age-group and then falls very sharply, showing that the roles of married women and mothers are still more important than professional roles in those countries. Italy, Greece and Portugal generally follow the same pattern, but the reduction in employment rate with increasing age after the 20–4 peak is much slower and less linked to marriage. But even when children are no longer dependent, women do not start to work outside the home again. In the Federal Republic of Germany, the UK and the Netherlands the statistical trends show that life is divided into three distinct phases: work, child-bearing and work. Involvement in work outside the home peaks in the 20–4 age-group, is followed by a reduction, although attenuated in the period favourable to child-bearing, and then rises again after 35. Going back to work often involves part-time employment. In Denmark, France and East Germany most women carry on working through the child-bearing years with only short periods of maternity leave.

However, the entry of women into active life outside the home has mainly taken place in the form of part-time work. According to data from an EC Labour Force Survey in 1987 (results reported in EUROSTAT, 1989b), 27.8 per cent of working women exploit the possibility of part-time work. This figure rises to 42.1 per cent in Denmark, 44.6 per cent in the UK, and reaches its maximum at 57.2 per cent in the Netherlands, while in Italy it is still low at 10.4 per cent. Part-time workers are overwhelmingly women, as much as 81.6 per cent, to the extent that it may be said that the part-time labour force is a female labour force. The growth of under-employment is very often a response to unemployment. For example, in the UK part-time working has not only been more common than in most other European countries, but has increased substantially as unemployment has risen. It seems that what might be called involuntary part-time working among women is on the increase.

The gap between employment rates and activity rates is widening, revealing a larger reserve of unabsorbed female labour which in practice amounts to unemployment. This implies that, if demand does not increase in the near future, the job market for women will soon encounter severe problems. The unemployment rate among women in the EC, which was less than the rate among males in 1970, has progressively increased and at 11.6 per cent exceeded the rate among males in 1987. However, the Community mean conceals very diverse situations. Very high levels are found in Spain (27.6 per cent) and Eire (19.1 per cent) and much lower levels in West Germany (7.8 per cent), Denmark (7.6 per cent) and Luxembourg (4.1 per cent). Unemployment particularly affects women under 25, and rates vary from 41.4 per cent in Italy to 7.7 per cent in West Germany. But it should also be remembered that women, more than men, form part of the 'underground', 'hidden' or 'black' economy, widespread in many countries and particularly in Italy (Bagnasco, 1981). This type of activity may favour quicker savings according to objectives which the entire family nucleus aims at (Angelini and Magni, 1984, p. 177).

### Forms of Female Employment

All the countries of the Community are currently in that phase of post-industrial transition which, according to Bell (1973), is

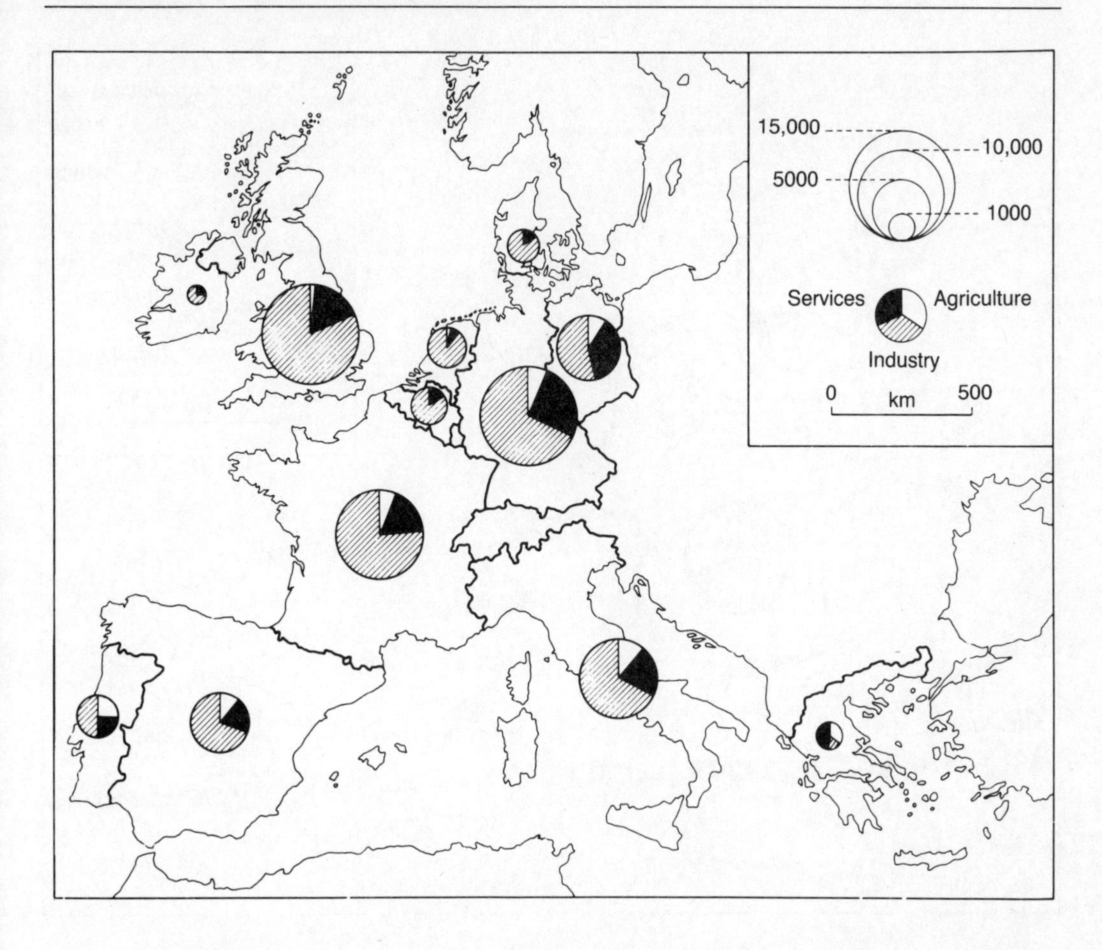


Figure 13.1 National variations in the percentage of women employed in three sectors of economic activity, 1986.

mainly characterized by a predominant services sector. Women, who have traditionally been concentrated in this sector, now take on a primary economic role. Their presence is particularly strong in the services sector (72.9 per cent of women are in this sector), rather weak in industry (19.8 per cent) and almost insignificant in agriculture (7.3 per cent). The advance of service employment has not been homogeneous throughout the EC, however. Over 80 per cent of women in employment are engaged in services sector jobs in the Benelux countries, Denmark and the UK, but in Portugal and Greece less than half are so employed (figure 13.1). In these two countries women are more frequently employed in the primary sector of the economy. In East Germany in the late 1980s, 52.7 per cent of women were employed in services, with 38.4 per cent in industry.

Within the three sectors the presence of women takes on varying and particular economic significance (figure 13.2). Women

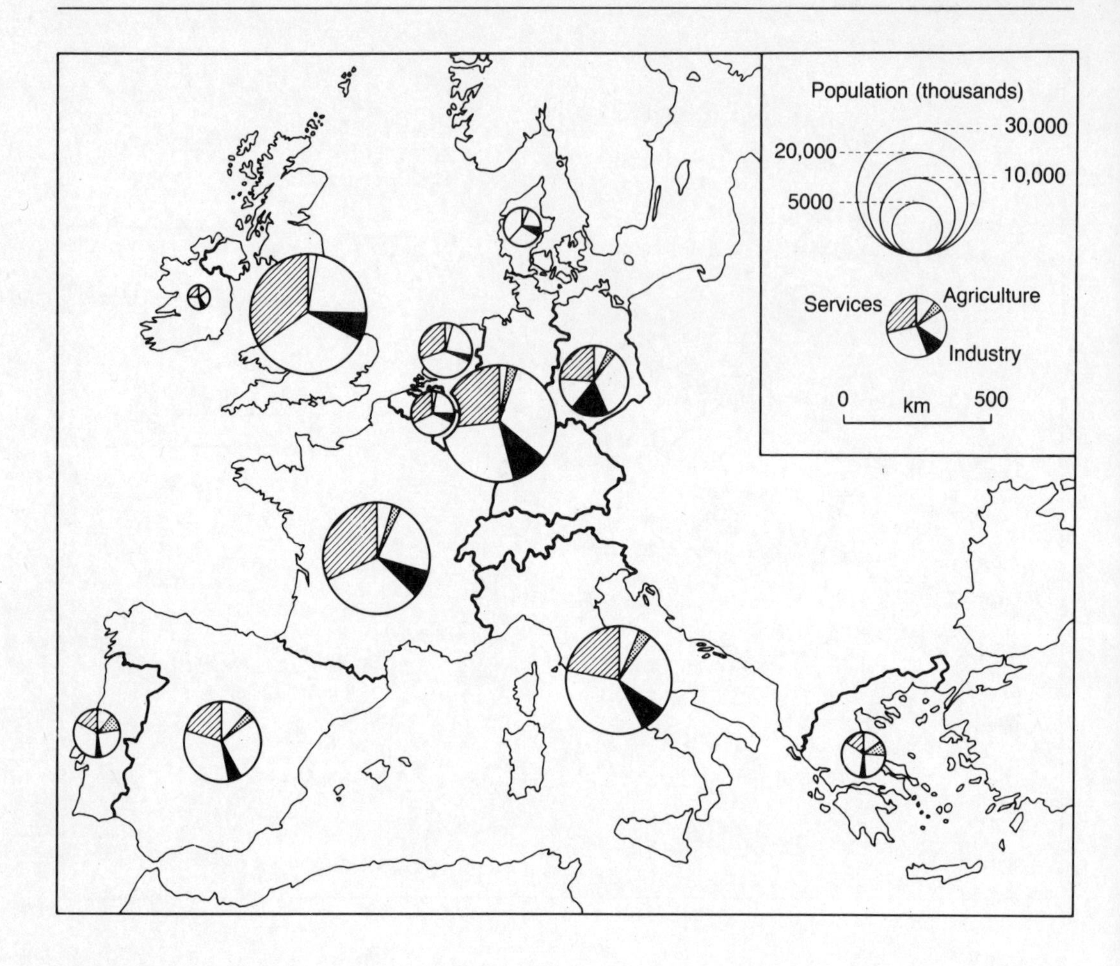


**Figure 13.2** National variations in quotas of women employed in three sectors of economic activity, 1986.

employees make up 46 per cent of the services sector, 35.3 per cent of the agricultural sector and only 23.3 per cent of the industrial sector. We may speak of 'feminization' of services, particularly in countries like Denmark and the UK, but in Portugal, Greece and West Germany women are still important in agriculture. In general, it is possible to recognize a process of segregation with men in the secondary and women in the primary and services sectors. However, within the secondary sector there are still some areas of specialization, such as textiles and clothing, that employ large numbers of women. But it is inside the services sector that the labour force is typically female, to the extent that one may speak of the 'female professions'. The most important examples are health, education, social assistance and social services; these are the 'human services' which, according to Bell (1973, p. 15), are typical of post-industrial society. In them, especially in the economically more advanced countries (Denmark, the UK, the Netherlands), the proportion that is female exceeds 70 per cent. In contrast, research and development, which may still be considered a central service in the current model of economic organization, contains relatively few women. The same may be said of business services, which has been expanding rapidly in recent years thanks to new technologies.

In terms of the place women occupy in their professional hierarchies, the position is usually inferior. In Denmark, a country with a large number of employed women, one in nine is self-employed compared with one in six men (Court, 1986, p. 100). Overall, women in 'top jobs' are very rare in the EC (Schaffer, 1981, p. 80).

The centrality of women's work and its current importance however, are threatened by the introduction of new technology, especially computerization, informatics, telematics and robotics. Much has been said and written in the last few years about the impact of new technologies on female labour (Bird, 1980; Gershuny, 1980; Werheke, 1983; Manacorda and Piva, 1985). It is generally argued that the introduction of these technologies will influence the transformation of work practices and opportunities for women. From the first alarming predictions (Bird, 1980), which envisaged a fall in the number of jobs for women ranging from 10 to 30 or 40 per cent in those sectors containing office, bank and insurance workers, it is now recognized that because of the speed with which innovation takes place no 'real-time' calculation of job loss is possible. On the other hand, the role of job loss must be considered in a wider context. In the current process of restructuring, although jobs are being lost, new ones are being established in the new industries that create the new technology. This is particularly true in the field of microelectronics. The focus must therefore be shifted towards supply, in the sense that it is the attitude of women towards work that must be modified; above all, more effort will need to be put into technical and scientific training for women. Although in all the countries of the Community education for women is apparently equivalent to that for men, most technical training remains male oriented.

Moreover, the new technologies are capable of opening up new prospects for women with 'telejobs', that is, work carried out far from the productive unit, mainly at home. The boom in the use of minicomputers and microcomputers means that going out to a fixed place of work may become increasingly less necessary. In this type of organization, women would find it easier to carry out their twofold role: productive and reproductive. However, this strategy has not been encouraged within the EC because it may involve the risk of social isolation. Instead, training is consistently viewed as the only strategy granting women a more important position in the job market. These considerations would appear to suggest that the greatest element of uncertainty regarding the future of the EC job market is linked to the development of the labour market for women.

### Working Women and Reproduction

The growth of women's work outside the home has coincided with the decline of fertility in all the countries of the EC, with the single exception of Eire, to the extent that the average number of children per woman has fallen to under 2.1, the replacement level. The situation suggests a fundamental question: is there a causal relationship between the two phenomena and, if so, how does it work? (See also chapters 5, 6 and 9.) The question is by no means straightforward and the various answers have been hotly debated, but as yet there is no convincing conclusion. Surveys carried out in several European countries, both at the aggregate level and by means of questionnaires, have shown that there is an inverse association between women's fertility and employment outside the home. A sample survey carried out in Italy in 1979, and subsequently incorporated in the World Fertility Survey (Salvini, 1984), provides an example. Table 13.2 shows how the average number of live births, which for the entire sample of 5499 married women aged 18–44 was 1.96, varied significantly with the women's employment situation. The lowest number of children were born to women who had full-time jobs outside the home; women who worked at home and those with part-time jobs had intermediate fertility levels; and full-time housewives were the most fertile, only seasonal workers being higher.

However, a causal relationship may not be inferred on the basis of this form of evidence alone. Fertility trends and those related to other characteristics of the job market may both depend on other variables. It may be that there is an interaction between the two phenomena or, as Schmid (1984, p. 18) suggests, that there is an 'ecological complex covering various relationships hidden within life-styles and professional situations'. However, it appears that cultural factors are now influencing fertility, while the influence of material factors, women's work being just one of them, is in sharp decline (Brunetta and Rotondi, 1989). More generally, it may be said that the causes of fertility decline in the second demographic revolution should be sought among the motivations of post-industrial society, in particular the increasing advance of individualism, with its destructive effects on traditional values such

**Table 13.2** Mean number of live births by actual work status, or work status in the past 12 months, Italian women, 1979

| *Work status* | *Women* | *Percentage* | *Mean number of live births* |
|---|---|---|---|
| Working women | | | |
| Full-time outside the home | 1482 | 27.0 | 1.54 |
| Part-time outside the home | 371 | 6.7 | 1.80 |
| Seasonal outside the home | 121 | 2.2 | 2.43 |
| Occasional outside the home | 102 | 1.8 | 2.21 |
| At home | 334 | 6.1 | 1.86 |
| At home and outside the home | 76 | 1.4 | 1.95 |
| Women not working | | | |
| Unemployed | 60 | 1.1 | 1.54 |
| Retired | 10 | 0.2 | 2.45 |
| Student | 16 | 0.3 | 1.24 |
| Looking for first job | 30 | 0.5 | 0.96 |
| Housewives | 2898 | 52.7 | 2.20 |
| Total | 5499 | 100.0 | 1.96 |

*Source*: Salvini, 1984

as marriage and parenthood. This has moved western European society from 'king-child with parents to king-pair with a child' (van de Kaa, 1987, p. 11).

## Concluding Remarks

Female labour has increased in all the countries of the EC, although women have not reached the same degree of employment as men. This increase has occurred mainly in the services sector of the economy and particularly in the so-called human services. However, female labour is often undertaken on a part-time basis or as part of the informal economy, and it is threatened by the introduction of new technologies. The role that women will play in the future will depend to a great extent on how the conflict between their two roles, production and reproduction, will be resolved. It may therefore be said that women's work outside the home has contributed towards hastening the fall in fertility, although this may also be due to the emergence of new cultural factors typical of post-industrial society.

# 14

# Internal Migration and Mobility

Gildas Simon

Whereas Europe, and in particular that part of it which is today encompassed by the European Community (EC), has for decades been the principal focus of emigration in the world, intra-Community and intra-European mobility is relatively low. At the present time, less than 2 per cent of Community nationals reside in a different member state, that is, about 5 million persons out of a total population of 323 million in 1987. This is very little compared with mobility between the states of the USA, but these differences are bound up with the history and the political structure of these two large economic entities that make simple comparison a difficult matter. It is not a great deal, especially in relation to the extra-Community mobility of the Community's population, since the number of EC emigrants living outside that entity is three times the corresponding figure for the interior, namely a total of 15 million, two-thirds of whom have settled, most of them permanently, in seven countries: the USA, Canada, Venezuela, Brazil, Argentina, South Africa and Australia (Baines, 1991). It is true that the days of mass emigration, which started in the mid-nineteenth century when emigrants left the continent in millions, have ended, but the attraction of overseas countries is still there (see chapter 2). No less than 600,000 EC nationals left their countries for the USA during the period 1970–80. This traditional attraction of lands beyond the seas, and in particular North America, for the skilled and highly skilled among the population remains an active component of the migratory equation within the EC.

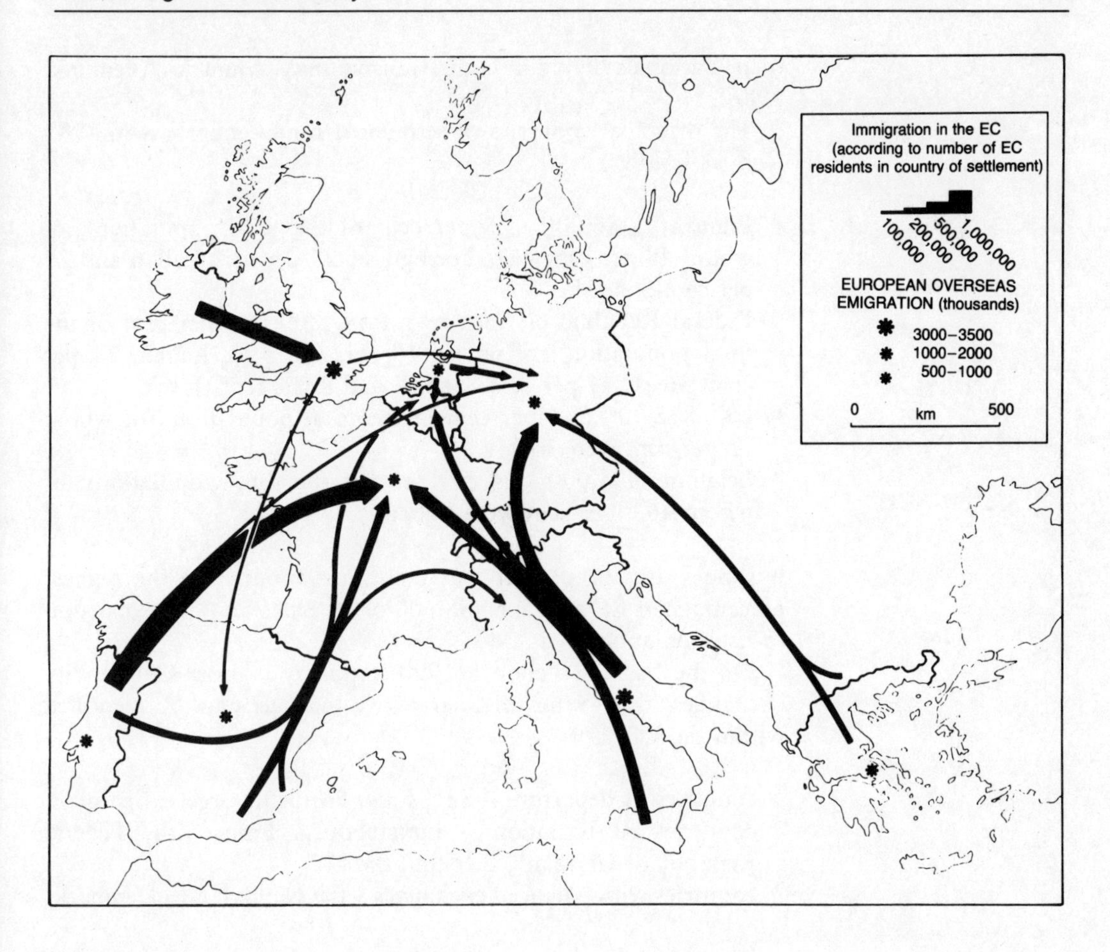


**Figure 14.1** Mobility of European Community nationals, 1988–90.

### From the Peripheral Regions to the Centre

The geographical aspect of intra-Community migration is characterized by a high degree of polarization (figure 14.1). Flows from the less-developed countries in the peripheral regions are towards the 'centre', the 'European megalopolis', the quadrilateral bounded by London, Lyons, Munich and Copenhagen. Nearly two-thirds (63 per cent) of EC immigrants are natives of the southern countries: Italy (1.2 million), Portugal (0.9 million), Spain (0.6 million) and Greece (0.4 million). In the west, Eire (0.5 million) and the peripheral regions of the UK (0.3 million) constitute a secondary region of high out-migration. On the basis of their populations, Eire and Portugal, both veteran lands for overseas emigration, are the two countries most involved in intra-Community emigration with 15.8 and 9.3 per cent respectively of their nationals living in other Community countries (Penninx, 1986).

The principal countries of settlement for EC migrants in 1987 were as follows.

- France: 1,386,000 (2.6 per cent of the total population), of whom 48 per cent were Portuguese, 21 per cent Italian and 20 per cent Spanish.
- Federal Republic of Germany: 1,437,000 (2.4 per cent of the total population), of whom 40 per cent were Italian, 21 per cent Greek, 11 per cent Spanish and 8 per cent Dutch.
- UK: 812,000 (1.4 per cent of the total population), of whom 72 per cent were Irish.
- Belgium: 495,000 (5.4 per cent of the total population), of whom 46.7 per cent were Italian.

It should also be pointed out that Luxembourg has the highest percentage of EC migrants (88,600 or 24.2 per cent) in proportion to its population.

On the basis of Lebon's (1990) typology of intra-Community exchanges, three types of migratory situations may be identified within the EC:

1 countries of departure – Eire, Italy, Portugal, Greece, Spain;
2 countries of reception – Luxembourg, France, the Federal Republic of Germany, Belgium, the UK;
3 countries with balanced exchanges – the Netherlands, Denmark.

These population movements have their foundations in history. For example, the inflow of Belgians, Italians and Spaniards into France, of Irish into Britain and of Portuguese into Spain are all of long standing. But in general the vast majority of EC migrants in the central area, and especially the continental area, of the Community arrived during the years 1955–73, a period of significant reconstruction and economic expansion (Salt and Clout, 1976; H. Werner, 1986; Straubhaar, 1988; van de Kaa, 1991).

The actual numbers which are made up of first-generation migrants and their descendants afford only a mere hint of the scale of mass migration that took place during the 1930s throughout north-west Europe. In the early 1970s the annual outflow to the north from countries in the south of the EC ranged from 500,000 to 600,000. The approximate breakdown is as follows: Italy, 200,000; Portugal, 170,000; Spain, 150,000; and Greece, 68,000. Many of these economic migrants only went to work in the north for a few years and then returned to their countries of origin. It was in the same period and for the same reasons that millions of workers from the Mediterranean countries, especially Yugoslavia, Turkey, Algeria, Morocco and Tunisia, arrived in the EC.

The reversal of migration policies in the countries of the north in 1973–4, for reasons just as political as economic, the halting of immigration of new workers and the introduction of incentives to return home slowed down considerably the flow from the external regions to the south. Nevertheless, immigration connected with family reunification has persisted, although at a generally decreasing pace. Italy alone has benefited from freedom of movement within the Community. The flow of return migrants to countries in the south of the EC has also been substantial.

In summary, during the last three decades the EC has developed a complex migratory system in which nearly all forms of internal and external mobility, intra- and extra-Community, are intermingled, not forgetting the ever-present considerable attraction of overseas countries. All these population exchanges – movements of seasonal and temporary workers with high or low qualifications, regular migrations by those who return to their own countries every year – and all these networks of connectivity form one of the most durable foundations for economic and human relations in the Community, and no doubt also one of the most reliable vectors of integration, which is also being brought about as part of ordinary everyday life.

### The Decline of Intra-Community Mobility

The fact that intra-Community mobility appears to have declined since the early 1970s may come as a surprise, especially when restrictions on the flow of people and goods are being removed, but the statistical indicators available to us tend to show a falling off of Community migration in most of the member countries. In the absence of continuous and comparable series for the past 20 years, it will only be possible to use rather crude indicators. One is the total number of non-national Community residents in the Federal Republic of Germany and France and a second relates to the labour force characteristics of certain EC countries.

In France and the Federal Republic of Germany, the two main host centres in terms of Community migration, the fall in numbers has been considerable. In France there were 1,860,000 in 1975, 1,577,000 in 1982 and 1,386,000 in 1987, while in the Federal Republic of Germany there were 1,616,000 in 1975, 1,496,000 in 1982 and 1,437,000 in 1987 (Bergues, 1973; Guillon, 1988). The trend seems to be even more pronounced among members of the economically active population. The number of Community workers in the Federal Republic fell from 732,000 in 1980 to 498,000 in 1986; in France from 653,000 in 1981 to 589,600 in 1986; in the UK from 406,300 in 1981 to 398,200 in 1985; in Belgium from 158,600 in 1980 to 140,700 in 1985; and in the Netherlands from 83,800 in 1980 to 76,200 in 1986.

Admittedly, the reversal of migration policies in the northern countries has undoubtedly affected emigration from the south: in Spain and Portugal, to which transitional arrangements continued to apply until 1993, and Greece, which was subject to this system until 1989. On the other hand, the return flows and the process of integration by the *jus soli* and by naturalization in the host countries account for the falling-off in numbers. The reverse of the coin is that other populations, non-Community citizens, who have frequently suffered more severely from the constraints of the new migration policy since 1973, have recorded a net increase in the EC. This applies to the North Africans who contributed about 1 million migrants to the EC in the early 1970s and 2 million today. It is also the case with the Turks whose numbers have increased from 650,000 in 1973 to the present figure of about 2 million.

Further, the decline of Irish emigration to Great Britain (452,000 economically active persons in 1978; 268,000 in 1986) and of Italian emigration (200,000 in 1973; 85,000 in 1980 of whom 64,000 stayed in Europe; 58,000 in 1986 of whom 44,000 stayed in Europe) provide good examples of the downward trend of intra-Community mobility. They also reflect the demise of that migration model which describes the first industrial era, one based on transfers of poorly qualified labour from the underdeveloped rural regions at the periphery to the industrialized local labour markets of the north.

## Migration and its Association with Regional Economic Disparities

There would appear to be an anomaly between the slowing down of mobility rates and the persistence of strong economic disparities in Europe if, that is, migration is thought of as a means of adjustment to disequilibria in terms of economic performance, unemployment and living standards.

It is known that in spite of an overall improvement in employment in the EC in the mid- to late-1980s, there are still marked contrasts in the breakdown between individual countries and in particular between regions in which levels of general unemployment, youth unemployment and long-term unemployment differ markedly (see chapters 11, 12 and especially 13).

To judge by the 1989 statistics, a distinction can be made between three types of countries according to their levels of total unemployment:

1 countries with relatively low unemployment – Luxembourg, Portugal, the Federal Republic of Germany, Denmark, Greece and the UK;
2 countries with average rates of unemployment – France, the Netherlands, Italy and Belgium;
3 countries with high unemployment – Eire and Spain.

A more realistic analysis based on youth unemployment, and therefore closer to the particular circumstances of migration, reveals a geography of under-employment marked by still greater contrasts (see figure 12.2). Here a distinction can be made between four types of countries:

1 countries with hardly any youth unemployment – Luxembourg and the Federal Republic of Germany for example;
2 countries with only low unemployment – the UK, Denmark, Portugal and the Netherlands;
3 countries with youth unemployment close to the EC average – Belgium and France;
4 countries in which from a quarter to a third of those aged under 25 are unemployed – Eire, Greece, Italy and Spain.

Gradients in unemployment rates are even steeper when viewed at the inter-regional level. There are potential flows of young persons from the peripheral regions, such as Extremadura, Andalucia, Sardinia and southern Italy, where nearly one in two young people is without work, to the major foci of economic growth in the Community's central area. In theory, therefore, we should be witnessing a reproduction of the migratory model that characterized the European labour market in the decade from 1960 to 1970; however, this system seems no longer to apply with any force.

Inequalities in the distribution of wealth between the member countries are substantial (figure 14.2). On the basis of per capita income purchasing power, the EC divides into two groups of countries. The 'rich' are, in descending order, Luxembourg, Denmark, the Federal Republic of Germany, France, the Netherlands, Italy, the UK and Belgium; and the 'poor' are Spain, Eire,

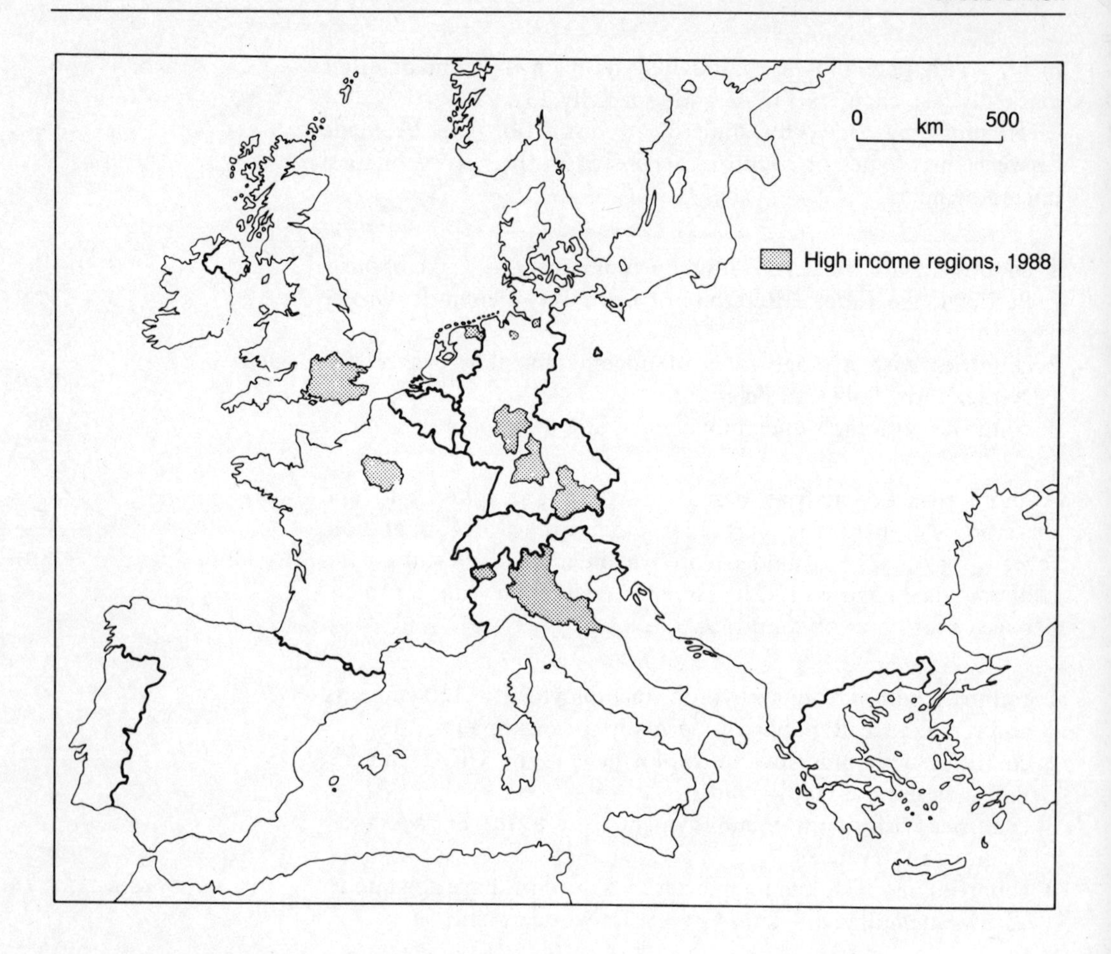


**Figure 14.2** Regions with the highest levels of purchasing power per inhabitant, 1988.

Greece and Portugal. It may be noted in passing that this rank order does not correspond exactly with that for employment. For instance, Portugal, which lies at the foot of the table as far as purchasing power is concerned, nevertheless has a relatively satisfactory employment situation. The disparity in purchasing power between Portugal and Luxembourg is about 2.5; between Greece and the Federal Republic of Germany it is about 2; and between Eire and the UK it is 1.7.

These contrasts, which are still more accentuated at the regional level, represent the principal factor in accounting for the organization of international fields of migration in western Europe from the outlying areas to the centre. This form of organization is a legacy of the past which has operated since the first industrial revolution and the concentration of capital in just a few favoured urban areas. But these considerable differences in living standards no longer seem to exert the same driving force.

**Restraints on and Barriers to Intra-Community Mobility**

The general falling-off of mobility in the EC is directly linked to demographic trends and the changing nature of the European economy. But there are, in addition, particular factors relating to the heterogeneity of economic, political and social experience within Europe.

The marked demographic contrasts between the south and the north and between countries and regions where the practice of religion is deep-seated and countries where it is not have been steadily narrowing since the early 1960s. The reduction of fertility has been particularly pronounced in the traditional emigration countries of the south. The decrease in fertility rates has been very rapid in Greece, Italy, Spain and, to a lesser extent, in Portugal. Today all these countries have a fertility rate less than that of France, which in the inter-war period was the archetypal country of very low fertility and immigration. Only Eire, whose emigration tradition is well known, has a birth rate that ensures the replacement of its population. This demographic convergence has gradually deprived the EC of one of the main engines of its migratory dynamism which for decades kept the countries in the heart of Europe supplied with population from the western and especially the southern periphery. The effects of this movement have been especially marked in Italy, Spain and Greece. It is no accident that these countries have in their turn become areas for immigration. The revolution in their demographic situation since the 1970s is one of the major explanatory factors. It is true that in the most outlying regions of the Community – northern Portugal, southern Spain, the Italian Mezzogiorno and the whole of Ireland – there are substantial reserves of young people who can be mobilized for emigration. But it is becoming increasingly probable that any available surpluses are going to be absorbed within the national entities where, as in Italy, strong inter-regional differences are being redressed.

The decline of intra-Community migration fits into the general context of the decline of internal mobility in Europe. The phenomenon is borne out in the majority of the largest and most populous member states: France, Spain and Italy (Tinacci Mossello, 1986; Rivière, 1987). In France, inter-regional exchanges have dwindled, sometimes appreciably, since 1975. More precisely, this falling-off affects the economically active population: between the two censuses of 1975 and 1982, less than 2 million economically active persons, that is, 8.45 per cent of the reference population, had settled in another region compared with 8.92 per cent in 1975.

The reasons for the all-round decline in general mobility are highly complex. They must no doubt be said to include the effect of the ageing process and the eradication of demographic disparities. They must also include the effect of the development of European societies and the overall improvement of living standards. Numerous factors hinder long-distance mobility: limitations on the transferability of housing and/or having a working spouse, for example. In short, mentalities are developing in the opposite direction to mobility, since in the Community the view that it is employment that should change its location towards the people and not the other way about is steadily gaining ground; this seems to be the embodiment of economic and social progress, induced in a certain sense by the ideology implicit in development within the Community setting. The potentialities for migration are for the most part encapsulated in the internal framework, where the continuous reorganization process involving the metropolitan regions and certain coastal areas is absorbing the human resources not only of the regions still affected by the rural exodus but also of the crisis-stricken areas. The more recent data on the Community area show that the effects of the crisis have disrupted the traditional migratory system; the migratory structures and attitudes of the years 1960–70 are rapidly disappearing.

Other factors arising from the EC's national divisions and cultural diversity appear to be slowing down or penalizing movement within the Community. The difficulties, even the current impossibility of adapting readily from one national system to another, and the fear of losing rights acquired in the country of origin or residence may constitute a serious barrier to the freedom of movement of workers within the Community.

The shortcomings of Community law in the field of social protection also pose problems for the right of abode of inactive persons in the EC. Under the present system, the member states will not be obliged to grant right of abode to all the Community's nationals where any of them go on social security. The issue is of particular concern to students and the elderly.

The role of cultural, and more particularly linguistic, factors in the mobility or immobility of the European populations must be considered carefully. On the one hand, it is true that not only the multiplicity of languages in the EC (no less than nine official languages) and the serious inadequacies as regards the learning and actual use of foreign languages but also the major differences in cultural customs and practices from one country to another are the real barriers to communication, to exchanges and, in the final analysis, to mobility. But on the other hand, the history of international migratory movements, and in particular those which have taken place in Europe, teaches us that total ignorance of a language or of the traditions and customs of another country has never stopped large-scale migration from developing if sufficiently powerful economic necessities make themselves felt, not only in the country of origin but also in the country of employment. The post-war migrations in which we have seen Italians and Spaniards moving in large numbers to work in the Federal Republic of Germany offer a good illustration of the way in which economic motivations outweigh language and cultural barriers. To stress the role of these cultural factors as the reason for the slackening-off of mobility in the EC would be to recognize implicitly that the economic disparities are narrowing appreciably in the Community.

It would appear that despite the differences and dissatisfactions, the gaps between costs and benefits are no longer sufficient to trigger off a mass-mobility process.

### The Development of Special Forms of Mobility

The prospects opened up by the total liberalization of movement for Community nationals in 1993 will doubtless not trigger off any shifts of population on a vast scale, both economically active and inactive, within the Community. The general ageing of European societies, the narrowing of the widest economic and social disparities, the inertia of cultural systems and the reformation of the national migration fields will tend towards a general stabilization of the Community's population. Portugal may prove the exception to this rule, for reasons both sociological – time-honoured migration traditions and the strength of a diaspora system – and strictly economic, such as the reserves of agricultural populations and the lag behind northern countries in terms of wage levels. However, it should not be concluded that there will be no movement at all after 1993. The new equation created by freedom of movement, the economic dynamism and the sociological context created by the completion of the internal market will create a new migratory equation in the area of the Community in which some of the most significant variables are already obvious. Three particular categories of population are most likely to be affected: young people, the highly skilled and certain categories of the non-EC population.

Young Europeans, and especially students, feel that the single market will provide greater opportunities for them to work abroad. But in the late 1980s only 2 per cent of students in the Community lived in countries other than their own. France and the Federal Republic of Germany were the leading host countries with nearly 20,000 EC students, while the two principal emigration countries were Italy and the Federal Republic of Germany again with 7000–10,000 students in other member states. The aim of the Erasmus programme is to attain a 10 per cent level of involvement of all students in the Community. The increase that may be expected to stem from these developments will contribute to improving the fluidity of the market for skilled employment, on which increasing pressures are now being exerted.

The concepts of 'skilled work' and 'job quality' are difficult to define with any precision and their development once the internal market has been completed are not easy to predict (Salt, 1988, 1992). However, the structural change in the economies, the technological trend and the development of tertiary activities suggest that undertakings will be seeking to recruit personnel with ever increasing levels of qualifications. A dearth of engineers and managerial staff is being reported in several countries, namely the UK, Germany and France, where the shortage of engineers is estimated at over 16,000 a year. In recent years, employment of managerial staff has shown a marked increase in Europe: in 1988, as many as 686,000 new posts were created in eight European countries, the majority by recruitment, the remainder by internal promotion. The services sector is ahead of manufacturing industry in terms of recruitment (159,000 against 137,000). The growing pressures in the market will stimulate the development of 'sophisticated migration' of skilled and highly skilled staff to the most active regions in the heart of the EC and to the expanding regions. This is likely to aggravate the existing imbalances in the distribution of technical competence and skills (for example, more than 50 per cent of French engineers live in the Paris region). But whether the mobility of managerial staff is spontaneous or organized within undertakings under the influence of the Europeanization of their corporate strategies it is likely that we shall see more and more short-term mobility – a few months to a few years – based on a system of shuttling between the country of employment and the country of residence. In short, in the case of technicians, engineers and managerial staff who will be promoted to move within the European area, the concept of movement will come increasingly to supersede that of migration.

However, even in this highly skilled European labour market, which should benefit from the increasingly favourable prospects for economic integration, there must be no under-estimating the trammels of mobility. Most European managers are highly selective as regards the areas in which they would prefer to work and demanding of rewards by way of compensation if their requirements cannot be met. In the professions, the total liberalization of services and the mutual recognition of qualifications will undoubtedly foster exchanges of human resources between countries, but these exchanges may be slow to take their full shape.

The case of the non-Community migrants is a complicated one which cannot be dealt with in any detail here (see chapters 15 and 17). Nearly 8 million nationals of non-Community countries are legally resident in the various EC countries. The Federal Republic of Germany and France between them contain two-thirds of the total: 3.2 million in the Federal Republic of Germany of whom 1.4 million are Turks; 2.1 million in France of whom about 1.4 million are from North Africa (figures 14.3 and 14.4). Even prior to 1993 strong circuits of mobility and information flow had developed among members of several of the larger ethnic groups involving extensive familial and commercial networks (Cohen, 1987). After 1993 these networks are likely to be strengthened and in some ways they provide a model for what may happen in the wider economy and society.

**Figure 14.3** Mobility of non-Community nationals within the European Community, 1988–90.

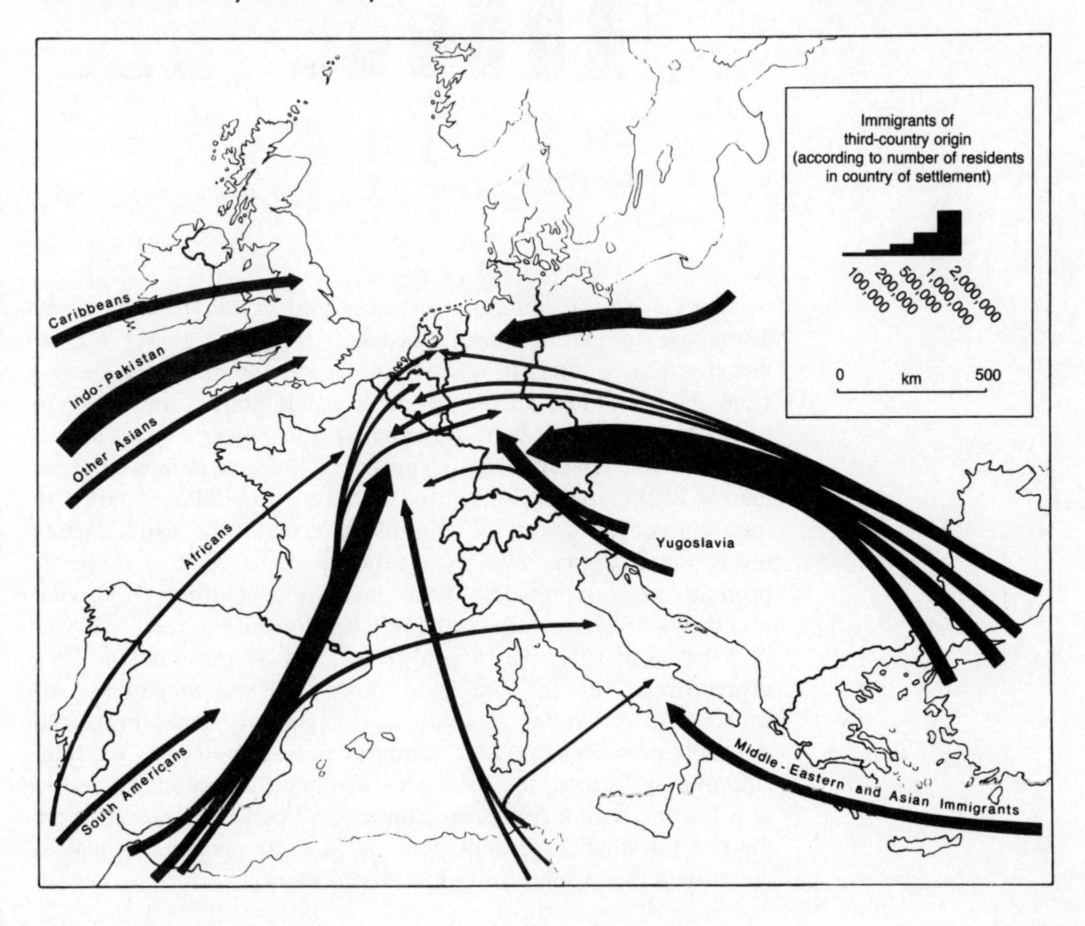

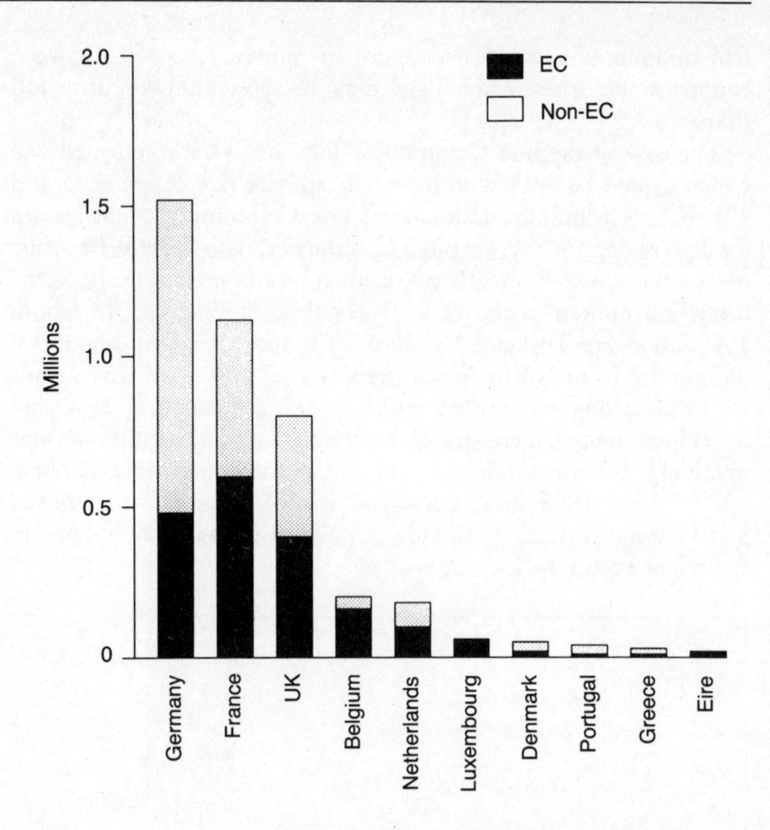


**Figure 14.4** Foreign workers in the member states of the European Community, 1986.

The clandestine migrants present quite a different problem. These are immigrants who have entered or who are residing illegally in a country. It is true that clandestine migration derives from the conditions in the countries of emigration and the representation of the country of employment which attracts migration, but the far-reaching dynamism of these movements is due mainly to the appeal exerted by the clandestine labour market in the immigration countries. The 'black' economy is also a fact of life in the EC, even if its significance in terms of gross domestic product varies appreciably from one state to another: 4 per cent was the estimate for France in 1989, but for Italy the estimate was 20 per cent in 1982. The appeal of this vast 'sector of activity' – a diversified sector and one with exceptional ramifications – its capacity to adapt to economic and regulatory conditions, and the unlimited demand for labour, even clandestine, in non-Community Europe, the Mediterranean and throughout the world in a Europe without physical frontiers, all point to an even more fluid clandestine movement potential and to the operation of a veritable underground European labour market.

But it would be unrealistic to dissociate totally the case of the clandestine migrants from that of the nationals of non-Community countries who have settled legally in the EC, even if their position *vis-à-vis* intra-Community mobility is different. The arrival and movement of the clandestine migrants in the Community are closely linked to the presence, dynamism and solidarity of the diasporas which have become established around immigrants who have legally settled in Europe. Such is the case with the Sri Lankans and the Pakistanis, the Vietnamese, the Chinese, the Lebanese, the Turks, the Tunisians and the Moroccans, to name only those national groups whose diaspora capacities, whether traditional or recent, are particularly manifest. The diasporas have become an essential participant in present and future mobility in the EC.

## Prospects for the Future

The prospects for mobility in the EC up to and into the twenty-first century are particularly difficult to discern. Crucial trends and virtually unpredictable changes constantly complicate the picture. The collapse of communism in the east and the reunification of Germany provide important examples of how political movements with far-reaching social and economic consequences can occur very rapidly and rather unexpectedly. These changes alone are likely to have very important implications for the wider pattern of European mobility in the future.

The Community, as has already been made clear at several points in this volume, is characterized by a pronounced imbalance in favour of the 'centre', that is, the south east of England, the Paris region, the Benelux countries and the German Rhineland. The powerful economic forces released into the large internal market and the increase in the power of Germany despite the cost of reunification cannot but strengthen the influence and therefore the attraction of the central area compared with the various peripheral regions. Will the Community's regional policy be sufficiently effective to prevent an aggravation of the differences in development between its wealthiest, the most active, regions and the less-favoured regions? In this respect the cases of Portugal and Eire will doubtless be particularly interesting. Apart from the 'centre' we may expect to see continued development along the 'sun belt', especially the Mediterranean coast: Europe's California where high-tech industries and the elderly may cluster.

On the demographic front, two crucial trends look as though they are likely to combine their effects. Within the Community, the mobility of the member states' national and foreign populations is in constant decline because of the ageing factor, which has not prevented the development of new forms of short-term movement. While migratory pressure is decreasing within the EC, it is constantly increasing outside, at the Community's frontiers. In the south, on the southern and western shores of the Mediterranean, where the demographic burden is increasing with the arrival of growing numbers of young people, many of them well educated, on the labour market, the perception of the disparities compared with the most highly developed countries and the effectiveness of the diasporas all must inevitably add to the pressure at the Community's frontiers. Moreover, the more or less mythical hope entertained by the migrant of benefiting from freedom of movement will give a still greater impulse to movement within the Community.

The recent upheavals in eastern Europe have not yet had their full effect on migration, even though recent developments in Germany provide a possible scenario. Highly skilled workers from eastern Europe may find a ready home in the west, but they may be accompanied by political and economic refugees whose status may remain under constant review and actual threat. Their arrival will be a matter for political debate and perhaps pressure from the wider international community.

# 15

# External International Migration

John Salt

Migrations into western Europe during the 1980s, particularly in the last few years, have been different in a number of respects from those of the guestworker days of the 1950s to 1970s (Castles et al., 1984). At that time there was more organization of movement, at least in the later years. A series of bilateral agreements facilitated the transfer of both information – on jobs and conditions in destinations and on labour at origins – and workers. Recruitment was often formally organized by both employers and governments: West Germany established its own recruitment offices in source countries, for example. With a regular turnover of labour between countries, if only on those returning home for holidays, a more or less organized system of flows developed.

It was not always like that. In the early days, during the 1950s and early 1960s especially, migrations were anything but organized. Seasonal agricultural workers drifted in and out of France, mainly from Iberia and Italy, without any real control and they were followed by construction labourers and then manufacturing workers. The original sectors for immigrant entry were characterized by small employers and informal organization. Only later was more rigid immigration control applied. Perhaps a quarter of migrant workers in the Federal Republic of Germany and three-quarters in France at the end of the 1960s were illegal (Böhning, 1972).

During the early guestworker period most migrant workers came from countries with long histories of international emigration: Italy, Spain, Portugal and Greece, in particular. There was thus an existing culture of migration that generated regular flows of reliable information. Later additions to the list of source countries were Yugoslavia and Turkey, countries not normally associated with international migration, and the flows from these areas differed in a number of respects from those emanating from the four countries already mentioned. There was a third group, the Maghreb; here the colonial relationship with France meant that there was a long-standing population flow in both directions, but movement to other parts of north-west Europe is a relatively recent phenomenon (Ogden and White, 1989) (see chapter 14).

Present day flows from the east and south of Europe combine aspects of all these. They also include large numbers of asylum seekers, some *bona fide*, some fraudulent; many of the latter would have been accepted as genuine refugees only a short time ago (International Migration Review, 1992). The new migrations from the east are often based on inadequate information on conditions in receiving countries, lack of preparation and exaggerated expectations. They come from countries which have experienced little emigration for decades, and so the flows are in many respects 'new'. In that so many seem speculative they are reminiscent of the moves out of the western Mediterranean lands in the 1950s and 1960s, and some evidence suggests that many of those eastern Europeans who get jobs do so in sectors of the economy similar to their Mediterranean predecessors, in agriculture, for example. Many, too, seem to be moving clandestinely, though it is doubtful if the term has the same meaning in today's context of tight border controls in western Europe as it did in the more relaxed conditions of 30 years ago.

Yet, in that most of the movers from the east have so far been Germans, either *Uberseidler* or *Ausseidler*, comparisons with the guestworker programme may not be valid (Castles, 1986; Frey, 1990). Since 1945 most foreign national migrant workers and their families have been entitled to temporary or denizen status, while ethnic Germans may become full German citizens immediately, with rights and privileges to match. In the structured labour markets of the west these new Germans have been reluctant to take the sorts of jobs offered or to work in the sugar-beet fields, like the Spaniards in de Gaulle's France. In any case the nature of the demand for labour has changed. No longer are western employers actively seeking to recruit migrant workers because of labour shortages, though the existence of niche employment should not be discounted.

Labour migration from the south has principally affected the former countries of emigration. Their growing prosperity has turned them into 'economic honeypots' in their own right, but there is considerable evidence to suggest that many migrants regard them as staging posts on the way to what they perceive as the lotus lands of the older industrial countries to the north. Many of these southern migrants are from the Maghreb, which has been exporting labour for decades. But people from the sub-Saharan countries have also been coming north in a resurgence of the movements of the 1960s and early 1970s that staffed the Parisian street cleansing department, for example. What is new for these migrations are the destinations rather than the origins and their unorganized and frequently illegal nature.

There are substantial differences between the economic and social contexts of the 1950s and 1960s compared with the 1970s and 1980s. The latter period has been one of economic crisis and the restructuring of production. Access channels to employment have changed, as have job requirements in the labour market. Different models are required to interpret flows today compared with what was seen to be appropriate 20 or 30 years ago (Rogers and Willekens, 1986). In the 1980s labour demand was often focused in sectors inaccessible to the mass of migrants, such as financial and health services and computing. There were jobs for migrants, but in marginal types of activity like fishing, agriculture, domestic service and catering. There was also the underground economy, usually low skilled, low paid and informal, but in its own way quite well organized, particularly in relation to ethnic recruitment networks.

### Data Problems

As international migration has come to occupy a higher position on the political agenda, the inadequate nature of the available database for informed debate has become all too obvious. Despite intense political interest, it is by no means clear that present flows exceed those of the past (see chapter 14). Around 1970, about a million migrants a year, mostly workers, moved into the principal immigration countries of north-west Europe, excluding the UK. In 1989 gross flows into the same countries were about 900,000, including ethnic Germans but excluding asylum seekers. What has changed is the increased numbers of asylum seekers, especially from Third World countries, and the increased inflows in recent years to Italy, Spain Portugal and Greece from southern Mediterranean countries.

The main problem in assessing how many international migrants there are at any one time in Europe, where they are moving from and to and who they are is the lack of accurate data and in some cases the complete absence of any data. By and large stock data come from censuses, but they are periodic, out of date by the time they become available and lack comparability between countries. Flow data are normally by-products of information collected for other purposes, from population registers or frontier controls. Even when one source is ostensibly common to a number of countries, critical differences may remain. The EC Labour Force Survey, for example, has a common core of questions and should produce comparable data, but its completion is voluntary and that affects its completion from one country to another. In the UK the response rate is about 80 per cent, but in the Netherlands it is not much more than 50 per cent and those least likely to answer are foreign immigrants.

Even where data are collected on a regular basis, the criteria used to define immigrants and emigrants vary between countries. In some countries an immigrant is someone who intends being a resident for more than three months (Belgium and Italy); elsewhere it is six (the Netherlands) or 12 months (UK and Eire); or there may be no defined period (Germany). Where side-by-side figures of migratory exchange between countries do exist, they rarely coincide. In 1987, out of a potential 132 pairs of data between European Community (EC) countries, there were only 55 possible comparisons: 29 of them showed differences of over 50 per cent and only 10 per cent were within the 10 per cent margin. So even where data exist and are published, they are not necessarily reliable.

For countries with less developed statistical services than those of the EC, and particularly for those states that have only recently become contributors to international migration, the degree of uncertainty is very substantial. Much of the current debate about actual and potential movement from the east and south – and especially back again, for emigration statistics are even more unreliable than those for immigration – is taking place in a data vacuum.

### Stocks of Foreign Population and Labour

It is not possible to provide accurate and comparable data across Europe on the stocks and flows of the foreign population because of differences in national definitions and sources. It is clear that during the 1980s stocks of foreign population have increased throughout western Europe. All countries for which data are available showed some increase between 1980 and 1989, and in some cases these increases were considerable, particular when set against increasing numbers of naturalizations which occurred during the same period (table 15.1). The Federal Republic of Germany experienced the largest absolute increase with a foreign

**Table 15.1** Available information on stocks of foreign populations in selected OECD countries, 1980–9 (1,000s)

| *Country* | *1980* | *1981* | *1982* | *1983* | *1984* | *1985* | *1986* | *1987* | *1988* | *1989* |
|---|---|---|---|---|---|---|---|---|---|---|
| Austria | 282.7 | 299.2 | 302.9 | 275.0 | 268.8 | 271.7 | 275.7 | 283.0 | 298.7 | 322.6 |
| Belgium | 878.6 | 885.7 | 891.2 | 890.9 | 897.6 | 846.5 | 853.2 | 862.5 | 868.8 | 880.8 |
| France | – | – | 3714.2 | – | – | 3752.2 | – | – | – | – |
| Germany (FRG) | 4453.3 | 4629.8 | 4666.9 | 4534.9 | 4363.7 | 4378.9 | 4512.7 | 4630.2 | 4489.1 | 4845.9 |
| Luxembourg | 94.3 | 95.4 | 95.6 | 96.2 | 96.9 | 98.0 | 96.8 | 98.6 | 100.9 | 104.0 |
| Netherlands | 520.9 | 537.6 | 546.5 | 552.4 | 558.7 | 552.5 | 568.0 | 591.8 | 623.7 | 641.9 |
| Norway | 82.6 | 86.5 | 90.6 | 94.7 | 97.8 | 101.5 | 109.3 | 123.7 | 135.9 | 140.3 |
| Sweden | 421.7 | 414.0 | 405.5 | 397.1 | 390.6 | 388.6 | 390.8 | 401.0 | 421.0 | 456.0 |
| Switzerland | 892.8 | 909.9 | 925.8 | 925.6 | 932.4 | 939.7 | 956.0 | 978.7 | 1006.5 | 1040.3 |
| UK | – | – | – | – | 1601.0 | 1730.0 | 1821.0 | 1839.0 | 1822.0 | 1948.0 |

born population of 4.846 million in 1989 compared with 4.453 million in 1980. Other large increases were to be found in Switzerland (plus 147,500) and the Netherlands (plus 121,000).

Despite the rising trend in migration, the distribution of the foreign population throughout Europe in 1989 was similar to that at the beginning of the decade, with West Germany and France accounting for the majority of Europe's foreign population (approximately 8.6 million in the two countries). The next largest stocks were in the UK and Switzerland with over 1 million foreigners, and Belgium with about 880,000.

The composition of the foreign born population in Europe in 1989 was a reflection of successive waves of post-war migration associated with labour shortage, and more recently, but especially since the mid-1970s, with family reunification. In most countries the foreign population has been dominated by a small number of national groups, sometimes a single one. In West Germany, for example, one-third of all foreign nationals in 1989 were Turks (over 1.6 million) and another 17 per cent were Yugoslavs. In France in 1985, 41 per cent of all foreign residents were nationals of either Algeria, Morocco or Tunisia (over 1.6 million people).

The existence of these dominant national groups has various implications. They provide policy-makers with large, defined target groups for integration measures, with a common language and culture. But their size and visibility may lead to resentment and even xenophobia among the indigenous population. In view of discussions within the EC about common policies on border control, the situation with regard to relative numbers of EC and non-EC nationals resident in member states is of considerable importance. In the Community as a whole in 1988, foreign nationals accounted for approximately 4 per cent of the resident population: EC foreigners made up 1.5 per cent and the non-EC foreigners 2.5 per cent (table 15.2). Of these 8 million foreign residents about 38 per cent were EC nationals; 14.2 per cent were nationals of one of the Maghreb countries; 15.3 per cent were Turks and 5.4 per cent were Yugoslavs. Hence, a majority of all foreigners were not EC nationals and therefore not subject to free movement provisions. Their numbers give some idea of the scale of the problem facing the Community in developing common policies towards third country nationals.

Despite general increases in the stocks of foreign population between 1980 and 1989, changes in the stocks of foreign labour have varied between the traditional countries of immigration (table 15.3). In 1989 the stock of foreign labour in West Germany (1.941 million) was 8.3 per cent lower than in 1980, despite an increase of 8.8 per cent in the foreign population. In Belgium and Switzerland, in contrast, stocks increased in each successive year after 1980.

The majority of foreign workers in Europe in 1989 – like the majority of the foreign population – was concentrated in West Germany and France, with a total of over 3.5 million workers. The UK had nearly another 1 million. The foreign labour stocks of each country reflect their respective populations of foreign nationals.

**Table 15.2** Composition of resident population in European Community countries by broad nationality breakdown, 1988 (per cent)

| *Country of residence* | *Non-EC foreigners* | *EC foreigners* | *Nationals* |
|---|---|---|---|
| Belgium | 3.3 | 5.4 | 91.3 |
| Denmark | 2.1 | 0.5 | 97.3 |
| Germany (FRG) | 5.2 | 2.1 | 92.7 |
| Greece | 1.1 | 1.1 | 97.8 |
| France | 3.8 | 2.8 | 93.4 |
| Italy | 0.6 | 0.2 | 99.3 |
| Eire | 0.5 | 1.9 | 97.7 |
| Luxembourg | 1.9 | 23.9 | 74.2 |
| Netherlands | 3.0 | 1.1 | 96.0 |
| Portugal | 0.7 | 0.2 | 99.1 |
| Spain | 0.4 | 0.5 | 99.1 |
| UK | 1.8 | 1.3 | 96.9 |
| Total | 2.5 | 1.5 | 96.1 |

*Source*: EUROSTAT, 1990

**Table 15.3** Available information on stocks of foreign labour in selected OECD countries, 1980–9 (1,000s)

| *Country* | *1980* | *1981* | *1982* | *1983* | *1984* | *1985* | *1986* | *1987* | *1988* | *1989* |
|---|---|---|---|---|---|---|---|---|---|---|
| Austria | 178.4 | 177.9 | 166.2 | 154.8 | 146.7 | 148.3 | 155.0 | 157.7 | 160.9 | 178.0 |
| Belgium | 332.7 | 332.2 | 338.9 | 375.0 | 388.3 | 396.3 | 403.1 | 411.5 | – | – |
| France | 1458.2 | 1427.1 | 1503.0 | 1557.5 | 1658.2 | 1649.2 | 1555.7 | 1524.9 | 1557.0 | 1593.8 |
| Germany (FRG) | 2115.7 | 2096.3 | 2029.0 | 1983.5 | 1854.9 | 1823.4 | 1833.7 | 1865.5 | 1910.6 | 1940.6 |
| Luxembourg | 51.9 | 52.2 | 52.3 | 53.8 | 53.0 | – | – | – | – | – |
| Netherlands | 188.1 | 192.7 | 185.2 | 173.7 | 168.8 | 165.8 | 169.0 | 175.7 | 176.0 | 192.0 |
| Sweden | 234.1 | 233.5 | 227.7 | 221.6 | 219.2 | 216.1 | 214.9 | 214.9 | 220.2 | 237.0 |
| Switzerland[a] | 501.2 | 515.1 | 526.2 | 529.8 | 539.3 | 549.3 | 566.9 | 587.7 | 607.8 | 631.8 |
| UK | – | – | – | – | 744.0 | 808.0 | 815.0 | 814.0 | 870.0 | 960.0 |

[a] Seasonal and frontier workers are not taken into account.

Despite provisions for the free movement of labour in the 12 member states of the Community, a majority of foreign workers continue to be non-Community nationals. About 46 per cent of foreign workers in the EC were member nationals in 1983 – by 1988 this figure had slipped slightly to 43.7 per cent – and 15 per cent were Turks, about 10 per cent were nationals of one of the Maghreb countries and 8.1 per cent were Yugoslavs.

#### Flows of Migrants

During the period 1960–89 the EC experienced a net migration of nearly 6.5 million, an annual average of about 214,000. From the mid-1980s the average annual net inflow was almost 500,000. Member states have been unevenly affected. During the 1980s well over 40 per cent of all net migration in the EC occurred in Germany, increasing to nearly three-quarters in the late 1980s. If the movements of ethnic Germans and asylum seekers are excluded, most immigration flows have been for purposes of family reunion: about 90 per cent of the gross inflow to Belgium and West Germany in 1988 and lower figures in France (70 per cent) and Switzerland (55 per cent), although still the majority. An upsurge in the entry of foreign workers has also occurred, especially in Austria, West Germany and the UK, but also in Switzerland, Belgium and France.

The causes of increased labour immigration are not clear. It has occurred despite continuing high levels of unemployment, especially among foreign workers. There is some evidence that labour inflows from outside the Community contain substantial numbers of the highly skilled. Certainly, between the industrialized countries there have been exchanges of professional, managerial and technical workers, many of them corporate transferees within the internal labour markets of transnational organizations. The importance of migration into the EC in general may be illustrated by the case of the UK. The vast majority of work permits, about four-fifths, have gone to professional and managerial workers. In recent years most long-term work permits for entry to the UK have gone to workers transferred by their employers.

### Refugees and Asylum Seekers

The marked escalation in asylum seeking in the 1980s has provided Europe with a new type of refugee problem. Existing humanitarian policies and procedures have proved inadequate to meet the task. While most applications for asylum in Germany have been made by those from eastern Europe, elsewhere in the Community applicants from the Third World are in the majority. In the UK the number of asylum seekers has risen dramatically, especially from Africa, prompting a new Asylum Bill in 1991.

A major problem has been created by the need to assess which of the claims are genuine. Between 50 and 60 per cent of claimants arrive without documents or with forgeries. After processing, about 75 per cent of applicants stay: about 50 per cent of them are legally accepted and the remaining 25 per cent stay on by other means. In Germany in 1990 only 4 per cent of asylum claims were judged to be genuine. Given this complexity, it is not surprising that increasing numbers of applicants have generated raised costs; the total for 11 European and North American receiving countries rose from under 1 billion dollars in 1983 to between 6 and 7 billion in 1990. Rising costs and sheer pressure of numbers now constitute major threats to the institution of asylum.

### Migrants from the East

Migration from eastern Europe and the former USSR presents a new challenge (Ruile, 1983; Gans, 1990). From the beginning of the 1970s to the mid-1980s, emigration from the Warsaw Pact countries was about 100,000 a year, half of whom were of German origin. In 1989, 1.2 million emigrated; about 720,000 were Germans, 320,000 were Turks from Bulgaria (many of whom subsequently returned) and 235,000 were Soviet citizens. In 1990 emigration from the USSR stood at 450,000. The political changes in the eastern bloc, signalled from 1985, have led to the creation of a new migration system.

Soviet emigration remains the great enigma (Vichnevsky and Zayontchkovskaia, 1991; Vichnevsky et al., 1991). Flows between the Commonwealth of Independent States and the west are influenced by changes in the attitudes of the various political regimes with respect to rights of exit. The number of short-term visits abroad has been increasing and since 1985 the number of former Soviet citizens emigrating has risen spectacularly. The figures for 1990 include approximately 200,000 Jews, 145,000 Germans, 50,000–60,000 Armenians and 20,000 Greeks. So far these outflows have affected only a relatively small number of countries: Israel, Germany, the USA, Greece and Canada.

Predictions of the likely scale of emigration from the CIS vary, the absence of data making the exercise even more a case of reading the runes than usual. Experts at the Moscow Academy of Sciences believe that annual emigration will peak at about 1.5 million in 1993. Subsequently, the number leaving is expected to be around half to three-quarters of a million a year. This is considerably less than some estimates that have come from the west. A survey of 3000 Soviet citizens in 1990 found that 16 per cent wanted to emigrate, but such exercises are notoriously uncertain predictors.

### Migration into Southern Europe

It is now clear that, with the exception of Germany, Europe's main destination countries are Greece, Italy, Spain and Portugal – all former countries of emigration (Koch, 1989). Economic growth, new social security systems and democratic and constitutional structures have combined to exert a strong pull. However, it is not clear to what extent they serve as transit states for migrants who wish to settle in the north west. These former countries of emigration in southern Europe have increased demand for unskilled labour owing to their expanding economies. At the same time, immigration restrictions and tighter border controls among northern European states have encouraged clandestine immigration from the developing countries into southern Europe. By the mid-1980s the economies of former countries of emigration in the south were experiencing labour shortages in areas unattractive to the indigenous population, especially among the service industries. The underground economies also began to flourish. The size of the foreign population in these countries is estimated at a minimum of 2.7 million and could well be as much as 3 million. Roughly half this number may be illegal immigrants. It has been estimated that at the present level of immigration the number of foreigners may reach about

5 million by the year 2000. Of these perhaps only 2 million will be legal migrants.

The origin countries are both in the Mediterranean region, notably the Maghreb, and further afield, although it must be said that there is a marked absence of accurate statistical data. Immigrants come mainly from Third World countries – Maghreb, Cape Verde, Philippines, Eritrea, Somalia, Jordan, Egypt, Latin America, Gambia, Ghana and Guinea. A Council of Europe report (1990a) estimated that, of 1,158,000 registered immigrants to Italy, Spain, Portugal and Greece, 25 per cent came from Third World countries. This percentage increases to 50 if illegal immigrants are taken into account.

The employment of most new immigrants in southern Europe tends to be on the margins of the labour market or, worse, in parallel markets largely uncontrolled and unsupported by a collective bargaining system (Golini et al., 1991). Informal recruitment channels operate in directing flows. The occupations taken by these new migrants include domestic service (especially among women), fishing, agriculture and mining, work in restaurants and hotels, and small industrial concerns. Many of the illegal immigrants are exploited, receiving low pay, experiencing poor living conditions and lacking union representation and social security benefits. Recruitment is frequently through ethnic networks, often to the embarrassment of the various governments concerned.

Spain has been at the forefront as a reception country for Europe's new immigrants (Muñoz-Perez and Izquierdo Escribano, 1989). The novelty of this situation is illustrated by the fact that until 1985 there was no legislation on the rights and freedoms of foreign nationals living there. Immigration grew steadily if unspectacularly during the first half of the 1980s, after which there was a dramatic increase in numbers, especially from Africa – Morocco, Guinea, the Gambia, Ghana and the Cape Verde Islands. The number of Asians also increased, especially from India, the Philippines and China.

Most of Italy's new migrants work in a range of service and marginal jobs, in terms of economic sector quite different from those occupied by Italians themselves in the north-west of Europe in the 1950s and 1960s (Gambino, 1984).

However, the characteristics of the main groups of immigrants differ. North Africans are mainly single men; they have tended to concentrate in the larger towns and cities and in the low grade tertiary jobs. Those from the Philippines are mainly women who often work as cleaners despite the fact that they are often well qualified and had skilled jobs in their home country. The same could also be said of the migrants from Iran and the political refugees from South America. But the immigrants from Sri Lanka are likely to be unqualified and may have entered Europe illegally. The group from sub-Saharan Africa, which is growing rapidly, has a particularly lowly employment record. The Chinese immigrants fall into two distinct groups: those who form part of the historic diaspora and who mostly engage in trade and craft industries, and more recent migrants from both the People's Republic of China and other parts of south-east Asia.

### Emigration Pressures from the South

Unlike the 1950s and 1960s when migration into north-west Europe was primarily driven by demand, present day mass movements stem mainly from excess supply. Decelerating population growth in the EC and European Free Trade Association countries coupled with continuing high rates of growth in the south have given the Mediterranean basin the steepest demographic gradient in the world. During the 1960s the ratio of population growth between the countries of the southern and eastern rim of the Mediterranean and those of north-west Europe was 3.3 to 1; it may be 17.6 to 1 in the 1990s (Golini et al., 1991).

Not only will differentials in the rates of population growth be accentuated, but many potential migrants from the southern Mediterranean will be well-educated urban dwellers who may become frustrated by the limited economic opportunities in their own countries. Demographic pressure from sheer numbers will be compounded by comparative age structures. In most European countries the ageing process is well under way (see chapter 8), in sharp contrast with the situation further south. These differences will have important implications for both immigration in the north, where there may be some labour shortages, and emigration from the south, where large numbers of school children will pass into the labour force and impose intolerable demands in terms of job creation. For example, over 40 per cent of the population of Algeria, Morocco, Tunisia and Egypt are currently aged under 15.

The longer-term demographic future may be even more disturbing. The population doubling time around the southern rim of the Mediterranean is 25 years. At that rate, by 2020 Algeria and Morocco may well have populations larger than that of the UK, while Egypt will be approaching the 100 million mark.

Sluggish economic performance seems likely to encourage northward migration given persisting economic differentials between Europe and especially North Africa. The export of labour from south to north could be one way to narrow the gap.

### Issues of Policy

Immigration policy in Europe has passed through a number of phases. Before 1973 the labour market dominated, with social and political considerations largely subservient to economic needs. Even before the oil crisis there was growing concern about the increasing size of the foreign population. This was made worse during the late 1970s by two factors. First, foreign nationals did not return home as labour market 'buffer' theories suggested they would. Second, and unexpectedly, they brought members of their families over to join them. What had been regarded as a temporary migration of workers was thus transformed into family settlement. A new phase of policy formulation was initiated in which questions of integration and assimilation dominated and in which the social issues of immigrant minorities often came to be classed with xenophobic reaction from host communities. During the 1980s this phase has overlapped with a new one, the development of policies to encourage return movement aimed at decreasing the size of the immigrant population. Aspects of this policy have ranged from little more than financial encouragement to a rather more studied attempt to assist returnees by the establishment of programmes involving training and aid in placement and investment (see chapter 16).

In the new immigration countries, especially Spain and Italy, these phases have been telescoped, though policies of repatriation have been replaced by regularization of illegal migrants. The result is that in western Europe as a whole there has been a general convergence of political objectives. All states would prefer less immigration (Coleman, 1992).

It is possible, therefore, to see the evolution of European migration policies in terms of a series of stages. First, up to and during the 1970s they were developed to cope with the economy's demand for labour, and were therefore generally characterized by a *laissez-faire* approach. Second, from the mid-1970s they have been based on the moral imperative to allow family reunion while at the same time limiting new, primary migration. Third, policies have been concerned with the need to integrate these new migrants more fully into economic, social and political life – thus far, largely unsuccessfully. Fourth, they are currently coping with 'push-pressure' movement from the south and east. Of particular concern are asylum seekers and illegal immigrants. The policy issues at the moment are mainly concerned with establishing entry controls. Finally, there is a continuing stage of policy harmonization. This is proceeding within the EC, but it also involves policy agreements between Community members, other European states (which might come under increasing immigration pressure should a 'fortress' mentality prevail) and the principal countries of emigration.

The future development of migration policy in Europe will depend upon how attitudes towards migrants evolve. It is probably true to say that Europe has evolved a generally negative ideological view of migration. International migrants are generally seen as a problem, in competition with native people for jobs, housing and so forth and the cause of a whole host of social problems. This is in contrast with the rather more positive view prevailing in the USA, where immigration is frequently seen as creating energetic communities, bringing innovation, enriching cultures and expanding the economy. Of course, this comparison is far too simple, but in most European countries migrants are viewed positively only as a temporary necessity. The nation-states of Europe are built round common concepts of culture, language and religion, into which new immigrants fit often uneasily.

# 16

# Migration Policies

Jürgen Bähr and Jörg Köhli

The migration flows into and between the countries of the European Community (EC) are regulated by two different types of legal measures. On the one hand, each of the member countries has its own migration policies and measures based on these policies, which apply primarily to citizens of non-Community countries. On the other hand, EC-wide agreements regulate the freedom and impose limitations on labour movements within the Community for EC citizens.

### Migration Policies of the European Community

In contrast with other categories of persons, since 1968 wage-earning workers from countries that are members of the EC have not been subject to any limitations whatsoever with respect to their workplace and thus residence (table 16.1). This principle continued to be upheld when Denmark, the UK and Eire joined in 1973. Transitional regulations were first applied when Greece became a member in 1981, and then Spain and Portugal in 1986, because it was feared that there would be too heavy an influx from these countries. The transition period for Greece ended on 1 January 1988; for Spain and Portugal the termination date was 1 January 1993 (H. Werner, 1986, p. 550; Penninx and Muus, 1989, p. 374).

From the experience to date we may conclude that the freedom of labour movement within the EC has led, and will continue to lead, to the creation of increasingly inter-linked migration flows, especially for highly trained employees, but we need not expect a

**Table 16.1** Regulations for European Community nationals settling in other European Community member states, by type of migrant

| | *Rights* | | |
|---|---|---|---|
| *Type of migrant* | *Initial settlement* | *Subsequent migration of dependants* | *Re-entry option (after temporary return to country of origin)* |
| Individual migrant workers | Free movement | Permitted | As for initial settlement |
| Company-linked migrant workers | Free movement | Permitted | As for initial settlement |
| Entrepreneurs | Free movement | Permitted | As for initial settlement |
| Students | No free movement[a] | Not permitted | As for initial settlement |
| Pensioners and persons with private incomes[b] | National legislation and regulations | | None[b] |
| Refugees | No free movement[c] | Not permitted | None |
| Illegals | No free movement | Not applicable | As for initial settlement |

[a] Except for children of legal immigrant workers.
[b] Residence as a retired pensioner after working in another Community country is as a rule permitted.
[c] Into member states other than that into which they are admitted.
*Source*: Penninx and Muus, 1989, p. 381

large-scale population shift from the less developed to the more highly developed regions of the Community. A large-scale influx of Italians, especially from the Mezzogiorno, did not take place after 1986, and the direction and volume of migration flows have not changed decisively since the UK joined the Community. Even migration from Greece towards the other EC countries has not been noticeably stronger since 1988. Rather, in many cases the countries of southern Europe have themselves developed into areas for illegal international migration, primarily from Africa and Asia, since the 1980s and have begun to apply stiffer migration controls (Koch, 1989; Muñoz-Perez and Izquierdo Escribano, 1989).

As yet the Community lacks any uniform policy on migration from non-EC countries, so that the laws and regulations of the separate member states are used to govern this area. Agreement exists only with regard to the right of re-entry. In none of the countries of the Community do nationals from non-EC countries normally have the right to re-enter their country of immigration once they have definitively left it (Penninx and Muus, 1989, p. 383).

A certain amount of standardization can be expected once the single European market has come into full effect. For example, the Schengen Agreement of June 1990 provided that from that time the border controls between the Federal Republic of Germany, Italy, France and the Benelux countries would be abolished and in 1991 Spain and Portugal also joined this agreement. As a logical consequence, a common external immigration policy was agreed toward the countries that were not party to the treaty (Martin et al., 1990, p. 602).

### Migration Policies of the Member States

Until the end of the 1960s or the beginning of the 1970s none of the EC countries had a real policy on migration and foreign workers; the predominant attitude was definitely *laissez faire*. At best the state created conditions making it as easy as possible for business and industry to import foreign workers via recruitment agreements, for example. Belgium and France, in particular, tolerated 'spontaneous' and often illegal immigration for a long time, and in some cases it was subsequently legalized (by France in 1981–2 for around 130,000 persons) (Manfrass, 1983, p. 166).

The turning point did not come until 1973–5 when all countries recruiting 'guestworkers' began to experience an economic downturn in the wake of the first oil crisis. Now, for the first time, governments began to intervene and to attempt to control migration more rigorously than they had previously done. Obvious starting points for this were to be found in limiting further migration and encouraging return migration (Castles, 1986; H. Werner, 1986; Frey and Lubinski, 1987; Frey, 1990).

#### *Limiting Immigration*

The minor economic slump of 1966–7 had already shown that the employment of foreign labour and, by association, migration did not react to varying economic circumstances with the degree of flexibility that had generally been anticipated. As the average length of stay increased and more and more foreign workers sent for their dependants, it became obvious that in a prolonged economic crisis, like the one looming in 1973, return migration would not be automatic. All significant immigration countries reacted to the reduced demand for labour by imposing restrictive *ad hoc* measures. Such measures affected the admission of new foreign workers as well as the uniting of families of foreign workers who were already legal residents of the country. For the most part, uniform conditions existed with regard to the immigration of new foreign workers. Since the 1970s strict regulations have limited, though not entirely barred, immigration of persons who are not subject to the EC provisions on freedom of labour movement or, in the case of Denmark, those of the Nordic states. The special regulations in force in France regarding migration from the former colonies, especially from Algeria, were also gradually reduced by the beginning of the 1970s in favour of uniform state controls (Manfrass, 1983, p. 164; Castles, 1986, p. 764).

The Federal Republic of Germany was the first country in the EC to introduce a general ban on the recruitment of new labour. This was done in November 1973 when the labour offices were instructed as a matter of course not to issue any more work permits to foreigners who had entered the Federal territory after November 1974. In France and Belgium too the further influx of foreign workers was officially stopped after July–August 1974; Luxembourg followed in March 1982, though a number of exceptions were allowed, especially highly skilled workers and labour in the hotel business. In the Netherlands there was no formal ban on labour recruitment, but visa requirements were extended, effectively allowing the authorities to stop the entry of foreign workers into the country. For a long time the UK remained a special case because until 1962 there were no restrictions on the entry of Commonwealth citizens. British immigration laws were again changed fundamentally in 1971 and since the beginning of 1973 citizens of Commonwealth countries are also required to have work and residence permits to enter the country. Simultaneously, their previous legal claim to British citizenship after five years' residence was cancelled.

By means of such legal strictures and associated measures, in particular stricter entry controls and turning back would-be migrants at the borders, the legal immigration of foreign workers was successfully reduced so that the discussion very soon turned to the question of whether and to what extent family reunification should also be restricted. Governments and social groups both agreed unanimously that complete entry prohibition for family members was out of the question for both legal and humanitarian reasons. The debate was therefore directed towards establishing an age limit for children and defining which dependants would have the right to join a foreign worker. The associated questions of the extent to which individuals who had themselves entered the host country as dependants were entitled to have their own dependants join them and whether dependants in general should have free access to the labour market were also raised.

As a rule all countries allow spouses and children who are minors to join workers who are already resident as long as they meet certain conditions. However, in many countries the age limit has been reduced. For example, in the Federal Republic of Germany it was reduced from 21 to 16–18 in 1981, depending on the region, and in Belgium from 21 to 18 in 1984. Further restrictions affect family reunification with second generation spouses, which has either been stopped completely or made conditional on a certain length of residence, duration of marriage or minimum income. In general, access to the labour market was also made more difficult for spouses and juveniles joining workers who were already resident. In the Netherlands the right to family reunification is handled comparatively liberally (Castles, 1986, p. 766; Penninx, 1986): the age limit for dependants is 21; restrictions apply mainly to spouses of second generation immigrants. Not least for this reason, the Netherlands, registered a considerable increase in the number of foreigners even in the 1980s, contrary to other western European countries. However, the regulations applied in the UK are particularly strict. As a rule a foreigner has a right to be joined by dependants only if he has a permanent residence permit, and even under these conditions the rights of husbands, in contrast to wives, are limited in order to prevent bogus marriages.

Since the mid-1980s the problems that a large foreign population creates have taken on a new dimension because of the greatly increased number of persons seeking asylum. Growing hostility toward foreigners has been the result, and in the end foreigners who have already been resident for many years have also suffered. France and the Federal Republic of Germany have been particularly affected. In France the annual number of applications for asylum rose from 20,000 at the beginning of the 1980s to more than 60,000 in 1989; in Germany it increased from around 30,000 to 120,000 with 193,000 in 1990. The persons seeking asylum no longer come from the traditonal 'guestworker countries', but increasingly from eastern and south-eastern Europe as well as various African countries. The majority are not victims of political persecution, but economic refugees. An indication of this is the low recognition rate: only 4.4 per cent in the Federal Republic of Germany in 1990. Even after their claims of persecution at home have been rejected, most of these foreigners remain in the country; deportation, however, is the exception rather than the rule, so that at the beginning of 1991 there were more than 1 million refugees in the Federal Republic of Germany. This situation not only encourages further migration, but it leads increasingly to domestic political problems which in the long run could undermine the right of asylum for persons who really are being persecuted. Similar restrictions have recently been introduced in Switzerland and Austria.

The recent changes in eastern Europe have radically affected the migration experience of certain parts of north-western Europe. Ethnic Germans living in eastern Europe are increasingly taking advantage of the easing of travel restrictions everywhere to be repatriated to the Federal Republic. A total of 2.4 million people were resettled in the period 1950–90. Of these, 780,000 alone came in the years 1989–90. The leading countries of origin in 1990 were the former USSR (37 per cent), Poland (34 per cent) and Romania (28 per cent). In these countries hundreds of thousands more have applied for exit permits and the German minority is still quite sizeable (around 2 million ethnic Germans in the CIS), so that these migration flows can be expected to continue (see chapter 17).

### *Encouraging Return Migration*

In view of the realization that restrictions on migration had at best helped to stabilize the number of foreigners but had not contributed to a lasting reduction, political interest increasingly turned toward encouraging foreign workers to return to their home countries. Two kinds of aid for potential return migrants can be distinguished. First, reintegration aid is intended to improve the conditions for returning foreign workers in their home countries. Such measures include loans for house construction or for starting new enterprises, customs exemptions, promotion of training programmes or development projects. Second, repatriation grant schemes or incentives consist of bonuses offered for return migration. This often involves paying out shares in insurance plans and reimbursement of travel expenses. Whereas the rationalization for the first category of measures, such as were already practised by the Federal Republic of Germany and the Netherlands in the 1970s, is primarily that they represent a kind of development aid, the second type more or less directly pursues the goal of lowering the numbers of foreigners by directly subsidizing return migration.

The French government was the first to begin an active policy of encouraging return migration, which it began early in 1975. However, the financial incentives adopted at that time were so low (on average only about 2000 francs), that they failed to have any great effect. Not until the bonuses were raised and made more accessible to a larger group of persons did the acceptance of the programme increase so that between June 1977 and December 1981 approximately 94,000 foreigners claimed their bonuses and left the country. Nevertheless, this can hardly be considered a success because the persons availing themselves of the programme presumably intended to leave the country anyway and were happy to 'take along a little extra incentive'. Besides, most of those involved did not belong to the category of foreigners that the French government most wanted to persuade to return home – North Africans. Rather they were migrants from Spain and Portugal which it was interested in having stay on (Manfrass, 1983, p. 167). For this reason this particular form of encouragement was discontinued in 1981.

Despite this lesson, the Federal Republic of Germany followed a similar course and passed a law encouraging return migration of foreign workers in November 1983. Under certain circumstances, foreign workers filing an application before July 1984 could claim a grant amounting to DM 10,500 plus DM 1500 for each child leaving the country with the applicant (H. Werner, 1986, p. 548). Furthermore, this law suspended the usual waiting period for refunding an employee's share in government old age pension schemes. Within the prescribed period 17,000 foreign workers applied for such grants. Simultaneously, around 125,000 foreigners had their contributions to the goverment old age pension schemes refunded. It has been estimated that around 300,000 foreigners, including family members, returned to their home countries because of the law encouraging return migration. Here too we may assume that the law primarily influenced the timing of return, but that it did not launch an entirely new trend.

Similar regulations were also enacted in 1985 in Belgium (until 1989) and the Netherlands and again in 1984 (with higher bonuses) in France. On the whole, the effects of these aid schemes on return migration and reintegration have not been very great, although many migrants have a latent desire to return home, as several surveys have shown. It is possible that more could have been achieved by raising the rather low subsidies and combining them with much more extensive aid for reintegration. However, this would have involved considerably higher costs.

The majority of the countries in western and central Europe will therefore be obliged to reconcile themselves to becoming countries of immigration with no appreciable reduction in the number of foreign residents in the foreseeable future. Indeed, it is more likely that they will have to expect a further influx. This will necessitate a fundamental change in policies towards immigrants and foreigners (Murray, 1989, p. 185). Restrictive measures alone will not suffice if some of the problems associated with having a large immigrant population are not to be exacerbated. What is needed is aid and encouragement to integrate. Apart from Sweden, the Netherlands has been most consistent in following this course and has openly accepted the concept of a multi-ethnic and multicultural society. Their new policies directed towards minorities aim to create a society in which such groups have an equal place and full opportunities to develop both individually and as groups. One result of this policy can be seen in an amendment to the Dutch constitution passed in February 1982 which gave foreigners the right to vote and to stand for office at the local level. Nevertheless, the Dutch are also aware that such a policy can only be successful in the long run if the number of foreigners does not continue to rise. Consequently, stiffer regulations on immigration and subsidies to encourage return migration are not seen as contradictory to the principle of integration; rather they are considered a pre-requisite for a sound immigration policy (Entzinger, 1985, p. 64).

Most other European countries are still reluctant to make fundamental changes to their policies towards foreigners, although everywhere there are unmistakable efforts to improve the status of foreign residents, especially of those who have been in the country for a long time, by granting permanent residence rights or easing naturalization procedures, for example. Particularly in France, the policy towards foreigners is about to reach a turning point. On the one hand, the classical concept, which was based on an unlimited willingness to assimilate foreigners, no longer works and is increasingly being questioned, especially by Islamic groups. On the other hand, in sections of the French population resistance to the creation of a multi-ethnic society is hardening. There is a distinct danger that with the creation of the single European market citizens of non-Community states will be relegate to the position of unqualified workers with little security as far as rights of residence are concerned. They will become second class citizens compared with members of the EC (Frey, 1990, p. 124).

# 17

# Ethnic Minority Communities in Europe

Paul White

Very few European countries are ethnically homogeneous in terms of population composition. Indeed, in many countries, such as Belgium, Switzerland and the old state of Yugoslavia, the degree of heterogeneity is a startling feature, contradicting as it does traditional nineteenth-century views on the definition of the 'nation' and the 'nation-state'.

Ethnic diversity can arise from a number of processes tied up with historico-political evolution and with economic growth and change. In the European context a distinction can be made between 'indigenous' minorities and those originating in recent population migration. The former groups inhabit territory that has been associated with them for, in most cases, many centuries but where that territory is now within the jurisdiction of a state of which the ethnic and cultural character is determined by another group. Such is the situation for the Basques in Spain, the Bretons and Corsicans in France, the Welsh in the UK, the Turks in Bulgaria and the Magyars in Romania. Such communities will not be considered further in this chapter, although many important demographic issues can arise in these cases.

Rather, the focus of this chapter is on the second type of minority situation identified above, of communities whose origins lie in recent migration into a nation-state where their ethnic character differs from that of the majority or of those groups that make up the hegemony of controlling interests. This distinction between 'indigenous' and 'migrant-origin' minorities is, of course, somewhat arbitrary since there are many cases that are difficult to classify. How long must a migrant-origin community, such as

Jews or Gypsies, be present for it to be labelled 'indigenous'? What of countries like Switzerland and Belgium where the political hegemony contains more than one ethnic group?

Definitional issues are, in fact, becoming more important and sensitive as time goes on. The very fact of continuing to identify certain individuals as belonging to a minority group of migrant origin tends to perpetuate an ideology of 'difference' and to encourage the continuance of stereotyping. This is particularly significant in the language used to describe certain groups, or the failure to identify others. In virtually all European languages a strong difference is implied by the use of the word 'immigrants' (French *immigrés*, Dutch *immigranten*, Spanish *inmigrantes*) rather than 'foreigners' (French *étrangers*, Dutch *vreemdelingen*, Spanish *extranjeros*). Terms in the first set are used to describe low-status ethnic minority communities while terms in the second set generally relate to higher-status groups, for example those drawn from the countries of western Europe with a similar level of economic development. German has become weighed down with the concept of 'guestworkers' (*gastarbeiter*) as the largest sub-group within the total population of 'foreigners' (*Ausländer*).

These semantic differences reflect certain public perceptions of a hierarchy of minority groups in terms of 'social difference' or 'acceptability' (Girard, 1977; Hagendoorn and Hraba, 1989); they also influence research and data collection. Thus the body of available research literature on 'immigrant' groups is considerably larger than the literature on other foreign groups. The Irish constitute one of the largest ethnic minority groups in the UK, yet in the series of reports on ethnic minority composition published by staff of the Office of Population Censuses and Surveys in Britain (Haskey, 1988b, 1989c, 1991; Shaw, 1988) they are completely ignored, as also are other 'white' minorities. On the other hand, in countries such as Switzerland and Belgium, where 'white' minorities have been regarded as low-status 'immigrants', considerable research data have been amassed (Bolzman et al., 1987; Diáz Alvarez, 1989). There are also dangers in the over-definition of minorities: thus one major British study (Bhat et al., 1988) starts by unapologetically labelling all non-white residents of the UK as 'black' – a label rejected by many South Asians and Chinese. One interesting piece of labelling concerns the identification of the *Aussiedler* moving to Germany from the former USSR, Poland, Romania and other parts of eastern Europe: 38 per cent of West German respondents to a recent opinion poll said these migrants were German while 36 per cent said they were foreign (Herdegen, 1989, pp. 332–3).

The failure to identify particular groups could be seen as

progressive, since continued labelling may act to retard the processes of integration. This is particularly important when consideration is given to those who have no direct experience of migration but who nevertheless are often seen as belonging to an ethnic minority community of migrant origin. The identification of this so-called 'second generation' is particularly sensitive. In many western European countries today a high proportion of 'ethnic minority' individuals have been born in the country of residence. In the UK it is estimated that in the mid-1980s 43 per cent of the ethnic minority population (using the definition outlined above) had been born there, rising to 53 per cent of the population of West Indian origin (Haskey, 1988b, p. 30). The question of self-identification among the 'second generation' – a term which Heather Booth (1985b) says will soon have outlived its usefulness – is a crucial issue.

Gonvers et al. (1981), in a study of young Spaniards in Switzerland, found that the high figure of 72 per cent expressed a strong desire to 'return' to Spain, but the definition of the second generation adopted in this study was such as to include those who had migrated from Spain as teenagers with their employed parents. In a recent study in the same country, Bolzman et al. (1987) suggest that the identity of the second generation is in part conditioned by their definitional status to the authorities. In France, where children born to foreign parents automatically gain French citizenship at the age of 18, they may prefer to use a holistic view of French society as a model for their own integration and when such integration does not occur they may take political action or develop alternative self-identification as a group neither French nor foreign, as is the case with the *beurs* (Hargreaves, 1989). In contrast, in Switzerland, where no such group naturalization exists, Bolzman et al. suggest that the children of immigrant parents can seek integration only through self-advancement, particularly in employment. However, as Glebe (1990, p. 266) points out, where ethnic minorities experience high degrees of residential segregation, the chance of inter-group contact is diminished and 'intra-ethnic peer group orientation' is likely to increase.

The fact that definitions and labels are social and political constructs is nowhere more clear than in examining the official bases for data collection on ethnic minority communities. The variety of these and the incompleteness of the data-collection exercises in many countries mean that it is now impossible to produce anything other than the most general international comparative statistical information. It is worth briefly reviewing the position in certain different countries.

The UK, from having some of the poorest statistical information in the early 1980s, with British passport-holding 'immigrants' in the decennial census only being 'correctly' identified by use of birthplace data or data on the birthplace of the head of the household (Booth, 1985a), has become one of Europe's most enlightened countries through the use of a self-assessment ethnicity question in the 1991 census (Ní Bhrolcháin, 1990). This is the only valid attempt to overcome the problems of the 'external' labelling of individuals as discussed above: nevertheless, 'white' minority groups are still ignored. During the 1980s increasing use was made of data derived from the Labour Force Survey in which a direct question has been asked on ethnic origin since 1979 (Shaw, 1988). However, such data sources are only of limited use except at the national level.

In the Netherlands, data collection occurs through a continuous population register, but this produces considerable problems in the identification of one very significant ethnic minority population – the Surinamese who predominantly hold Dutch citizenship, having arrived before their country gained its independence in 1975 (van Amersfoort and Cortie, 1973, p. 283; Tas, 1987). On the other hand van Amersfoort has also suggested that because of the diversity of ethnic and class backgrounds of the Surinamese it is not plausible to regard them as being one minority community (van Amersfoort, 1982, p. 173). In general, Dutch definitions revolve around nationality data, but children born in the Netherlands to a mother also born in the country have automatic Dutch citizenship, so that an inter-generational citizenship drift is inevitable. That is not the case in Germany or in Switzerland where gaining naturalization is very difficult. In Sweden, naturalization is very common, such that in 1982 there were 310,000 foreign-born 'Swedes' to add to the 405,000 foreign citizens in the country (Booth, 1985b).

French data, unlike those of the countries just mentioned, normally come from the population census and are based on nationality – a feature that obscures the fact that 'foreign' children born in France become French at the age of 18 unless they revoke that right. France also has the Dutch and British problem of ethnic minority passport holders from overseas territories (Butcher and Ogden, 1984), in this case still a politically integrated part of France. In addition, the relatively lax application of French immigration laws over the decades means that there are considerable numbers of clandestine migrants in the country. Guillon (1988, p. 30) has estimated that in 1982 the true number of foreign citizens in the country was possibly 8 per cent higher than the census figure.

Finally, the problem of clandestine migration becomes most acute when southern Europe is considered. In both Italy and Spain ethnic minority communities (the phrase itself has, as yet, no real conceptual significance) are at present equated with the number of foreign citizens. Figures for these are woefully inadequate. Collicelli and di Cori (1986) quote estimates as high as 1.4 million for the number of clandestine migrants in Italy in the mid-1980s. It is not yet known how many clandestine migrants came forward to make their status legal during the armistice granted in March–April 1991, a repetition of the tactic used in 1986 when 103,000 came forward, although this was thought to be a small proportion of the number eligible (Barsotti and Lecchini, 1989, p. 48). Spain had a similar armistice for regularization during 1985–6, with similarly disappointing results (Izquierdo Escribano, 1990). Recent estimates suggest there are around 170,000 clandestine migrants in Spain, which would add a further 30 per cent to the size of the known foreigner population. Attempts are currently being made in both Italy and Spain to control the in-migrant flows more strictly and to improve data collection methods concerning foreigners already in the country (Gozálvez Pérez, 1990).

The net result of the issues discussed so far is that in the early 1990s we are rapidly reaching the position where the production of aggregate scientific statements about the demographic characteristics of ethnic minority populations in Europe is becoming impossible. First, issues of definition are becoming ever more important, with the dangers of imposed 'external' definitions contradicting the more subtle processes of self-identification and integration of groups and individuals occurring year by year. Second, and partly following on from these difficulties of definition, the accuracy of the data that are available must be called into question. Whilst table 17.1 gives the latest official figures on foreign populations in various European countries, it must be stressed that these figures are crude statements of a much more complex reality since there is no exact equation linking foreign citizenship in official figures to the sizes of ethnic minority communities. Nevertheless, it is on the basis of such official data that most demographic research must be undertaken.

But the character of ethnic minority communities is constantly changing in western Europe. The remainder of this chapter considers some of the processes leading to such changes as well as some of the major issues for the future.

## Processes

Every community is in a state of constant adjustment to the forces acting upon it, both from outside and from within. During the 1980s a number of changes occurred in the processes affecting ethnic minority communities in Europe. These changes in certain

**Table 17.1** Foreigner and ethnic minority populations in selected western European countries

| *Country* | *Year* | *Foreign population total* | *Foreigners as percentage of total population* | *Notes* |
|---|---|---|---|---|
| Austria | 1989 | 303,500 | 4.0 | |
| Belgium | 1988 | 859,200 | 8.7 | |
| Denmark | 1989 | 143,600 | 2.8 | |
| France | 1989 | 3,679,500 | 6.6 | Plus possibly 300,000 unregistered foreigners and over 1.4 million (1982) naturalized foreigner-born French |
| Germany (FRG) | 1987 | 4,654,400 | 7.6 | |
| Greece | 1988 | 219,800 | 2.2 | |
| Iceland | 1988 | 4,800 | 1.9 | |
| Italy | 1986 | 450,200 | 0.8 | Current (1991) estimates total 1.35 million (2.5%) including clandestines |
| Luxembourg | 1987 | 96,400 | 25.9 | |
| Netherlands | 1988 | 588,600 | 4.0 | Plus possibly 190,000 Dutch citizens of Surinamese origin |
| Norway | 1989 | 136,000 | 3.2 | |
| Portugal | 1987 | 92,400 | 0.9 | Plus clandestines |
| Spain | 1989 | 311,300 | 0.8 | Plus clandestines |
| Sweden | 1988 | 401,000 | 4.8 | Plus over 300,000 naturalized foreigner-born Swedes (Booth, 1985b) |
| Switzerland | 1988 | 1,017,900 | 15.5 | |
| UK | 1987 | 1,767,100 | 3.1 | Total non-white ethnic minority population estimated (Haskey, 1991) as 4.7% or approximately 2.65 million |

*Source*: Council of Europe, 1990b

cases had their antecedents in the transformation of international labour migration flows during the 1970s, relating both to changing macro-scale economic circumstances and to social and political developments within individual countries (White and Woods, 1983; White, 1986). In total six processes now stand out as conditioning the present and future evolution of ethnic minority communities in western Europe.

### *Political Processes*

It is useful to distinguish between migration policies which affect international movement and community policies which affect resident minorities of migrant origin (see chapter 16). These two sets of policies do not necessarily act harmoniously together – indeed in some countries in recent years they have tended to contradict each other. For example, the aims of the dual policy in West Germany have been to foster ethnic minorities' options and desires to return 'home' whilst at the same time improving their living conditions in Germany, which could serve to encourage them to 'stay' (Glebe, 1990, p. 260). Arguably, it is only in Sweden and France that policy frameworks have partially evolved to reflect a concept of these as countries of immigration (Heisler, 1986), but even in France the operation of policy differs according to the races involved. All other western European countries still regard themselves as countries of emigration.

Throughout western Europe policies on migration became restrictive during the 1970s, or earlier in the case of the UK, and have often been tightened up still further during the 1980s, for example by restrictions on in-migration of certain categories of dependants (Glebe, 1990, pp. 260–1) or by visa requirements for entrants from certain countries (van Amersfoort and Surie, 1987, p. 176). Elsewhere, as in France, changes of policy in relation to immigration have reflected changes of the political complexion of government. Only Switzerland has retained its immigration policy structure unchanged throughout the last two decades (Bertrand, 1983).

However, it is important to note here the provisions for free movement of labour within both the European Community (EC) and the Nordic Community. One of the biggest challenges for the EC during the 1990s will be to combine immigration policies with the ideal of free movement within the Community. This problem is particularly exacerbated by the fact that clandestine entry into certain EC countries is relatively easy, and once some form of regularization has been undertaken the clandestine migrant can then gain rights permitting secondary migration to another country, under an EC directive of 1985. Italy is the country whose borders and antiquated administrative systems (Gambino, 1984, pp. 178–9) appear to the rest of the EC as the most permeable in this respect. In April 1991 Spain acted to tighten up the regulation of immigration from the Maghreb. The other major problem for the EC immigration policy lies in the migratory desires of the peoples of eastern Europe.

Both France and West Germany have made attempts to encourage return migration through the offer of financial inducements, a policy that has been criticized as pandering to the view that any 'foreigner problem' can be solved by a reduction in their numbers (Gans, 1990, p. 28). In France the groups who took the greatest advantage of these incentives were those who were seen by the indigenous population as 'least different' (Italians and Spanish) and who were anyway more susceptible to return movement, rather than the more mistrusted Algerians. In West Germany the policy was only applicable to nationalities from outside the EC (P.N. Jones, 1990a, b, p. 225).

Policies towards minority communities have shown greater differences between countries. As already noted, the case of naturalization varies markedly from state to state. The involvement of state organizations, both national and local, in housing provision specifically for ethnic minority groups is also variable, sometimes within individual countries between different municipalities, as is the case in the Netherlands (van Amersfoort, 1982). As recession deepened during the 1980s policies which were once relatively liberal have been watered down, for example in West Germany (P.N. Jones, 1990b, p. 224).

### Family Reunification

After the general halting of large-scale international labour migration in the mid-1970s, processes of family reunification came to the fore as prime causes of the net migration inflows that continued to occur into most west European countries. Such family flows are still of considerable importance. For example, the migration of dependants made up 75 per cent of all New Commonwealth movement to the UK in 1989. Proportions of foreigners not living in families have generally fallen during the last 15 years (Lichtenberger, 1984, p. 111; van Hoorn, 1987, p. 41), but since 1980 the pace of decline has slowed as the vast majority of those who arrived in north-west Europe as guestworkers now have their families with them (van Amersfoort, 1987, p. 93). It is, of course, this recent process of family reunification among some of the larger minority groups that has produced the real transformation from 'immigrant workers' to 'ethnic minority communities'. A notable social dimension of the recency of reunification remains, however, the fact that significant proportions of young people have spent considerable proportions of their lives apart from one or both of their parents in fragmented family and household arrangements (Gonvers et al., 1981, p. 282). The reunification of families has considerable implications for housing demand and thus for factors of residential location.

### Internal Community Processes

Ethnic minority communities have been created through changes in the patterns of international migration flow and above all through family reunification. The social and cultural structure of these communities and the aspirations of the individuals and sub-groups making them up have then evolved in ways which reflect the interplay of wider social processes in the countries concerned with the norms of attitude and behaviour brought from elsewhere as part of the baggage of the original migrants. The extent to which ethnic minority communities inhabit two worlds, create their own world, or move back and forth between two worlds, even on a daily basis, is open to debate and interpretation (Förderzentrum Jugend Schreibt, 1980; Lichtenberger, 1984). The other set of questions that arises concerns the degree of assimilation or of integration undergone by the minority group, or whether some other social model of adaptation is more appropriate.

The operation of internal community processes is of considerable importance when considered in an inter-generational perspective. The work of King et al. (1989) on the Irish in Coventry suggests that this community now lacks a strong ethnic identity and is becoming largely indistinguishable from the rest of the population. However, it has taken over 40 years to reach this position and the Irish were, from the start, a much more 'invisible' minority than are many to be found elsewhere in western Europe today. In the Paris area Schnapper (1976) found that an analogous assimilation process had occurred for Italian migrants of less than 30 years' standing, despite the internal linguistic differences from the indigenous French. A very large number of indicators might potentially be used to identify changes in the behaviour of individuals along a continuum from perpetual reference to 'home' or original society through to total identification with place of current residence (Ruile, 1983; Lichtenberger, 1984).

Where minority communities have been present for less time, or take their behavioural norms from more 'different' sources, internal processes commonly pull in opposite directions at once – reaffirming distinctiveness but also promoting social change and economic advance as a goal. The tension between these two forces can be very strong, for example between parents with experience of migration and their children born in the country of residence. If parents hold strongly to the 'myth of return' they may discourage long periods of education for their children if they see no long-term future for them before re-migration (Glebe, 1990, p. 261). Hargreaves (1989, p. 89) has pointed out that in many respects minority community children can only achieve success in the wider world by reneging on the cultural heritage of their parents. Muxel (1988) has considered whether, where young people of North African family origins have grown up alongside French youths in the Paris region, there is a sharing of all attitudes; in fact even in this situation cultural background shows up as important, also creating particular gender differences within the ethnic minority community. It has been a common observation in France that young people from minority communities are striving to establish their own separate identity, partly in response to perceived external disapproval. A similar observation is possible in the UK and to some extent elsewhere (Gonvers et al., 1981, p. 291).

The position of women within ethnic minority community structures is important, particularly in relation to demographic behaviour. The fact that significant proportions of western Europe's ethnic minorities are Islamic creates the potential for strong differences between women's roles within these groups and those applying in the secularized world at large. The likely extent of change in gender differentials of behaviour in ethnic minority communities is very hard to predict at present.

### *Geographical Processes*

The creation of ethnic minority communities has been influenced by a series of geographical factors; prime amongst them is the extent of segregation, normally defined in residential terms. Such segregation helps to create the conditions for the emergence of strong community identity, reduces the likelihood of contact between a minority community and the surrounding society and provides a hearth for the socialization of succeeding generations. Questions of residential concentration and change are therefore very important in understanding minority community evolution, but the relevance of prevailing American models of such evolution has been queried repeatedly (O'Loughlin, 1980).

The evidence on segregation processes in western Europe clearly emphasizes the role of housing markets. In Düsseldorf the segregation of almost all *gastarbeiter* groups has increased during the 1980s (Glebe, 1990). The widespread increasing concentration of ethnic minorities in West Germany can be ascribed to the tighter housing market conditions of the 1980s in which the halting of social housing construction severely limited opportunities for the poor and hence for the *gastarbeiter* minorities (Nebe, 1988; Gans, 1990; Glebe, 1990).

The increasing numbers of ethnic minority families in the later 1970s and 1980s put great pressures on the social housing stock in many European countries, and that led, where such housing was made available, to a rapid process of suburbanization of minority communities, for example in France and in the Netherlands (van Hoorn, 1987; van Amersfoort, 1987, pp. 100–5).

However, there are two dangers in over-generalizing any discussion of the operation of the housing market. First, the balance of housing classes, the degree of access of ethnic minorities to social housing and the posibilities of owner-occupation amongst ethnic minority individuals have been variable between countries, and sometimes between cities, and through time. Thus, while in France in 1970 'immigrants' were practically excluded from social housing (Blanc, 1983, p. 135), by 1982 almost 24 per cent of 'foreigner-headed' households were living in such accommodation (White, 1987). Second, as the British experience of the renting Afro-Caribbean in contrast to the owner-occupying Pakistani and Indian minority groups shows, the interest in different types of tenure arrangements varies markedly between minority communities of different cultural backgrounds. Hence, the segregation of different ethnic minority groups from each other is generally of significance (Lichtenberger, 1984, p. 316).

Different residential distributions also exist at the regional scale, reflecting evolutionary processes of migratory flow and distributional forces created by economic conditions at different times. In the UK, West Indians are the biggest group in Inner London, whereas Indians are the biggest group in the West Midlands and Pakistanis are the biggest group in Yorkshire and Humberside (Haskey, 1991, p. 27). In West Germany, minority communities were established according to a northward diffusion process (Giese, 1978) so that the most recent arrivals (Turks) are concentrated in the north while earlier arrivals (such as the Italians) predominate in the south (P.N. Jones, 1990b, pp. 226–7).

### *Refugee Flows*

A significantly enhanced process affecting the evolution of western European ethnic minority communities during the 1980s has been refugee flow. Certainly earlier periods have been affected by movements such as the westward flow of the *Volkdeutschen* after the Second World War, repatriation of the Dutch from Indonesia and the French from Algeria and the forced migration of East African Asians to Britain.

Increased refugee flow during the 1980s has been a result of increased regional conflict on the international stage, which has been largely unaffected by the recent period of super-power co-operation. However, while a considerable proportion of the refugees and asylum-seekers in western Europe have certainly been subject to political persecution as defined under international conventions, it is also the case that a considerable 'refugee' demand for asylum in western European countries is probably a hidden result of the imposition of stricter controls on 'normal' immigration (Norro, 1990, p. 192). Thus from 20,000 refugees arriving at the borders of the EC in 1970 the number had increased to 250,000 in 1989. Outside the EC, Switzerland and Sweden have been traditionally liberal in handling asylum applications (White and Kesteloot, 1990).

One aspect of policy that has been attempted with refugees and should have had significant effects on community formation has related to locations of resettlement. In Britain and France with the Vietnamese boat-people (Robinson and Hale, 1990) and in Belgium for all refugees (Norro, 1990) policies of dispersion have been pursued. However, in both Britain and France the aim of creating small communities of Vietnamese refugees in towns with low proportions of ethnic minority residents has failed, largely through inadequate central government involvement. Instead, significant communities of these refugees have been established in distinctive neighbourhoods in the larger cities. In Belgium the dispersion policy has only been in operation since 1987, but although the number of municipalities with resident refugees rose from 25 per cent in 1986 to 48 per cent by 1989, the greatest concentrations still occur in the Brussels area (Norro, 1990, pp. 200–4). During the 1980s the German Federal government operated a dispersal policy for the *Aussiedler* from eastern Europe and the *Zuwanderer* from East Germany involving housing quotas for each Federal *land* (Sprink and Hellmann, 1989, p. 326).

### *Economic Restructuring*

Economic restructuring in many parts of Europe has had particular effects on ethnic minority communities. The growth of migration among the highly skilled between western European countries stemming from the increased internationalization of economic and control structures and from the tertiarization and quaternarization of the economy have already been considered in chapter 15. Communities of highly skilled foreigners are now established in many European cities, but with high degrees of circulation and rotation such that the composition of the groups is constantly changing (Glebe, 1986; White, 1988).

However, other aspects of economic change are associated with new migration flows and community growth elsewhere. Several observers of the rapid growth of foreigner groups in Italy have drawn attention to the correlation of large-scale immigration not just with the closing of opportunities in other countries in the mid-1970s (Mingione, 1985), but also with the major crisis of the

Italian labour market during the period 1968–74 and its reconstitution on new post-Fordist lines (Gambino, 1984). This break has been linked to urban–rural shifts in population growth and industrial location (Adamo, 1986), but has also fostered the development of the informal economy. Labour restructuring occurred with the state remaining silent, and the same state inactivity has been a notable feature in the face of large-scale immigration, predominantly from the Third World. As Barsotti and Lecchini (1989, p. 50) have observed, 'Third World immigration in Italy is thus without doubt post-industrial in nature.' The decentralization of production and immigration have gone hand in hand (Gambino, 1984, p. 175) and the informal tertiary sector employment has added to opportunities. It is possible that one-half of all Third World migrants have found employment in some form of domestic service (Barsotti and Lecchini, 1989, pp. 54–5), with employment of women particularly common. Whether true ethnic minority 'communities' are developing in these circumstances remains open to question. In Italy, the exploitation and isolation of foreign-born women servants who live-in with their employers limits the possibility of community contact developing (Arena, 1982).

Superficially, much employment of the clandestine migrants in the new informal sector in both Italy and elsewhere appears completely unorganized. In fact complex networks are often present in the spheres of migration initiation, employment and residence. Ethnographic research in other European countries has shown the high degree of organization involved and the extent of state or hegemonic complicity in such communities as Egyptian newspaper sellers in Vienna (Leitner, 1990) or West African pedlars (Meyze and Rose, 1983) and Mauritian street porters in Paris (Vuddamalay, 1990). Economic restructuring and clandestine or unregulated international migration have often come together to create new types of employment associated with particular ethnic minority groups.

## Demographic Variables

The first section of this chapter explained why it is difficult to be precise about the true size of ethnic minority communities in most west European countries and why the very act of defining such minority groups is today fraught with difficulty. Nevertheless, in any consideration of the evolution of such communities the presentation of demographic rates for fertility, mortality, nuptiality and migration is highly desirable in order to assess, albeit without full precision, the significance of the various processes at work. In order to calculate such rates some base population totals and characteristics have to be accepted, such as those presented in table 17.1. Here comment will be confined to apparent rates of population growth or decline rather than to absolute numbers.

Commentators in the 1980s showed, for a number of west European countries, how the earlier post-war ethnic minority communities – often composed of Europeans such as the Italians or Spanish – were stagnating or declining in the countries of north-west Europe, with often rapid growth concentrated in newer migratory groups such as those from North Africa or Turkey (Castles et al., 1984; Lebon, 1986). A feature of the 1980s appears to have been a marked reduction in the rate of growth of the total ethnic minority population in many countries (excepting those of southern Europe which have experienced rapid increases), but with continued changes in ethnic composition within these totals. The processes involved have already been considered in this chapter.

For example, in the Netherlands there has been a marked slowing of the rate of increase of the Moroccan and Turkish communities since 1980, largely brought about by the effective completion of family reunification by that date (van Amersfoort, 1987, p. 93). The number of foreigners in West Germany was actually in decline for much of the 1980s. Here the greatest fall were of Portuguese, Italians and Spanish, with lower declines among the Greeks, Yugoslavs and Turks. In the case of the Turks, substantial return migration was partly compensated by some continued family reunification movement and by a relatively high rate of natural increase (P.N. Jones, 1990a, b). Return migration of Spaniards from Belgium has continued during the 1980s (Diáz Alvarez, 1989).

One of the few countries in which the proportionate contribution of different ethnic minority groups to the total did not change greatly during the 1980s was the UK, although there was some reduction in the West Indian group brought about by return migration, compensated by growth in the number of Bangladeshis through family reunification (Shaw, 1988; Haskey, 1991). However, these conclusions relate only to non-white ethnicities; international passenger flow data suggest a growth in the size of otherwise undocumented groups drawn from other countries of the EC, such flows almost certainly involving the movement of highly skilled personnel (Bulusu, 1991).

One marked consequence of the differential growth and decline of individual ethnic minority groups is the ageing of some communities compared with the youthfulness of others. These contrasts are often further accentuated through differential fertility such that certain groups have high rates of natural increase whereas for others fertility is at or below the level of the majority population.

In general, the ethnic minority groups of western Europe can be divided into two categories. The first consists of those minorities who were established as full communities, involving family composition, before the mid-1970s. These communities are now showing a combination of relative population ageing, certainly in comparison with communities of the second group, low fertility rates and out-migration as some of the original migrants return to their places of origin.

In the second category are minorities in which family reunification has only occurred, or has continued to occur, over the last 15 years, and which tend to record high fertility rates, youthful populations, some element of continued sex imbalance and incipient growth possibilities.

Among the ethnic minorities of the first type in France are some of the long-established groups, such as the Italians and Spaniards, where the communities are distinctly 'middle-aged' and diminishing in total size (Wihtol de Wenden, 1985). The same applies to West Indians in the UK (Richmond, 1987, p. 138). Fertility is generally at the level of the indigenous population. Among women with Spanish nationality living in France, completed family size is now lower than for the French (Sporton, 1990; see also chapter 5), and in Luxembourg, where the minority populations are generally of European origin, the fertility of foreigners has been lower than that of the Luxembourg citizens since 1986.

In the second category come the Islamic populations of most countries. In some cases, such as France or the UK, family reunification has a long history, but initially fertility rates were high by European standards and the degree of cultural retention has often been great to the extent that demographic assimilation processes have been operating only slowly. One of the few studies to consider inter-generational differences in fertility in Islamic communities has shown that whereas amongst Algerian migrant women now living in France the average number of children born by the age of 25 was 1.61, for the next generation – women with Algerian nationality born in France who had declined to take up French citizenship – the figure by the same age was 1.02 (Sporton, 1990, p. 198). Similar reductions appear to have occurred among Turkish women in West Germany with the result that the contribution of foreigner births to all births in the Federal Republic fell from 18 per cent in 1975 to 11.4 per cent in 1984 (Tribalat, 1987, p. 370). The pace of fertility decline is a crucial element in ethnic minority community evolution, but is largely conjectural (van Amersfoort and Surie, 1987, p. 178). In Amsterdam fertility rates, although higher than among the Dutch, are now falling among the Turkish and Moroccan populations, but the age structure of these communities is dominated by young people, suggesting that fast population growth is still likely for some years to come (Dielman et al., 1991). One aberrant group in the Netherlands appears to be the Surinamese, among whom there was an increase in total fertility rate between 1977 and 1985 (Tas, 1987, pp. 77–8). The explanation for this is almost certainly to be found in the changing social class and ethnic composition of the Surinamese over this period when migration from Surinam was considerable.

The diversity of demographic structures between different ethnic groups seen in France is also reproduced elsewhere. In West Germany by the mid-1980s the fertility levels of the main foreigner nationalities were little above that of the Germans, the only exceptional group being the Turks (Tribalat, 1987, p. 378). In the UK many demographic and household variables are similar for the West Indian, African, Chinese, Arab and white populations identified in the Labour Force Survey, but the Pakistanis, Indians and Bangladeshis are very different, particularly with regard to variables related to family and household sizes. However, one distinguishing feature of the West Indians is the very high proportion of single-parent families (Haskey, 1989c). Illegitimacy rates for births to Surinamese women in the Netherlands were over five times the level for the entire Dutch population (Tas, 1987, p. 79).

Accurate mortality levels among ethnic minority populations are often difficult to determine since high proportions of older people are first generation migrants who return to their place of origin before death, such that crude death rates may appear extraordinarily low (Bähr and Gans, 1987, p. 96). Where more sophisticated data are available they tend to show standardized mortality rates above the levels applying to the entire population (Tas, 1987), just as morbidity rates are also above average (Bergues, 1973; Curtis and Ogden, 1986).

A demographic indicator of great sensitivity for future ethnic community development is the rate of inter-marriage. Values generally differ for males and females, and also vary between ethnic groups. In France in the early 1980s mixed marriages were very common indeed for Italians and Spaniards of both sexes, but low for Portuguese and Algerians. The proportion of male Algerians marrying French women was particularly low in view of both the size of the community and the length of its residence in France. For all nationalities except Spaniards mixed marriage rates were lower for females than for males (Muñoz-Perez and Tribalat, 1984). In the UK, West Indians were much more likely to have a 'white' partner than were Indians or Pakistanis, and Bangladeshis were the most likely of all to marry within their own community (Coleman, 1985). Among all groups, except the Chinese, it was men who were more likely to be in mixed marriages than women (Shaw, 1988, p. 7).

One final fascinating line of enquiry drawn from data on mixed marriages concerns the 'labelling' of the ethnicity of the children. Coleman (1985) and Shaw (1988) both use data from the UK Labour Force Survey in which respondents identify their own ethnicity and that of their household members. It is possible to calculate that the probability of a 'West Indian' male marrying a 'white' woman and then labelling their children 'white' is 0.06, producing a significant degree of inter-generational re-labelling. The probability of an Indian woman having a 'white' husband and labelling their children 'white' is less than 0.01. Assimilationist labelling tendencies are clearly highly differential between groups and genders.

The final set of demographic rates are those relating to migration. Here the concern is not with international movement but with internal movement within countries of residence. Reference has already been made to the ways in which refugees in various countries, although initially dispersed, have made secondary moves to become more concentrated in residential location. As the studies reported in Glebe and O'Loughlin (1987) make clear, such residential concentration is the norm for almost all ethnic minority groups in western Europe.

In recent years work in countries with population registers has tended to point to higher levels of ethnic minority residential mobility than among the majority population, but it appears, at least from German evidence, that this distinction is being lost especially as far as intra-urban mobility is concerned. Whereas in the past foreigners, particularly *gastarbeiter*, moved frequently in the local housing market in an attempt to secure better accommodation and more secure tenure, the tightening of the market during the 1980s has reduced opportunities and thus mobility and has tended to stabilize residential concentrations (Glebe, 1990). Even where such a reduction in overall mobility has not occurred, moves are becoming much more local than was the case in the past and are often contained within sub-areas (Gans, 1990). One of the main causes of longer-distance moves within urban areas – allocation to peripheral social housing – is no longer operative. There is also evidence of reductions in the levels of inter-regional mobility among ethnic minority groups during the 1980s, for example in the Netherlands (Atzema and Buursink, 1985, p. 39).

It might appear that the recent fall in the mobility rates of ethnic minority populations indicates a convergence of migration behaviour with the majority population. That is not the case. Instead, the ethnic minorities appear to be becoming more segregated with movement concentrated in a limited geographical, economic and housing space and with little interaction with the migration patterns of wider society. However, these suggestions are probably not applicable to southern Europe where the apparently high circulation levels of recent immigrants is reminiscent of patterns in northern Europe during the 1960s.

### Future Prospects

A commentator writing in 1981 would have been remarkably prescient if she or he had foreseen several important developments which have since affected ethnic minorities. The scale of refugee flows, the massive Third World migration into Italy and to a lesser extent Spain, the rapidity of the decline of fertility among many established minorities and the reduction of internal mobility rates were all at an embryonic stage or had scarcely been observable in the data available at that time. Other developments of the last decade, most notably the political transformation of eastern Europe, were completely unexpected. Prediction is always a hazardous task. Nevertheless, forecasts can be made about the future evolution of certain aspects of ethnic communities, even though there are very substantial margins for error as new and unforeseen forces come into play.

There will certainly be pressures towards further massive movement into western Europe, coming principally from two sources – the Third World, and particularly Africa, and eastern Europe and the former Soviet Union. Many of the movements from the latter source may actually reduce the ethnic minority populations of the world if they consist of, for example, those of German culture moving into Germany or Magyar minorities in Romania resettling in Hungary, but the possible movement west of large numbers of Poles and Russians will possibly rebuild older minority communities dating from the early years of the twentieth century and which have now become all but invisible.

In the case of movement from Africa, Collicelli and di Cori (1986) have shown that, while western Europe's population will probably only grow by 3.2 per cent between 1985 and 2000, increase in the rest of the Mediterranean Basin will probably be by over 44 per cent, creating in the Mediterranean Sea a replica of the Rio Grande that divides wealthy America from impoverished Mexico and which it is the ambition of many from the south to cross. Indeed, the Spanish already use the word *mojadas* to refer to illegal immigrants. Whatever the official policies adopted towards trans-frontier migrants, it is difficult to conceive of Europe in the year 2001 without new minority communities, possibly of clandestine migrants, in the course of establishment.

In terms of those communities already in existence, figure 17.1 clarifies the various mechanisms leading to growth or decline and the influence behind those mechanisms. With new immigration now all but stopped for many minority groups it will be the course of fertility that will continue to determine growth potential. Immigration has been influenced in the past by forces external to the community, government policy and economic conditions for example, and by the needs of the community, for instance the desire for family reunification. In the future the second of these influences will be of lesser significance now that family reunification is effectively complete for most groups.

Demographic factors reducing the size of minority communities are of three types. International out-migration, generally in the form of return migration, is controlled by governmental policy and by internal forces within minority communities. Over the next decade it might be expected that the arrival at retirement age of many of the first waves of migrants of the 1950s and 1960s might lead to a further acceleration of the patterns of return migration already observed (Wihtol de Wenden, 1985). An ageing minority

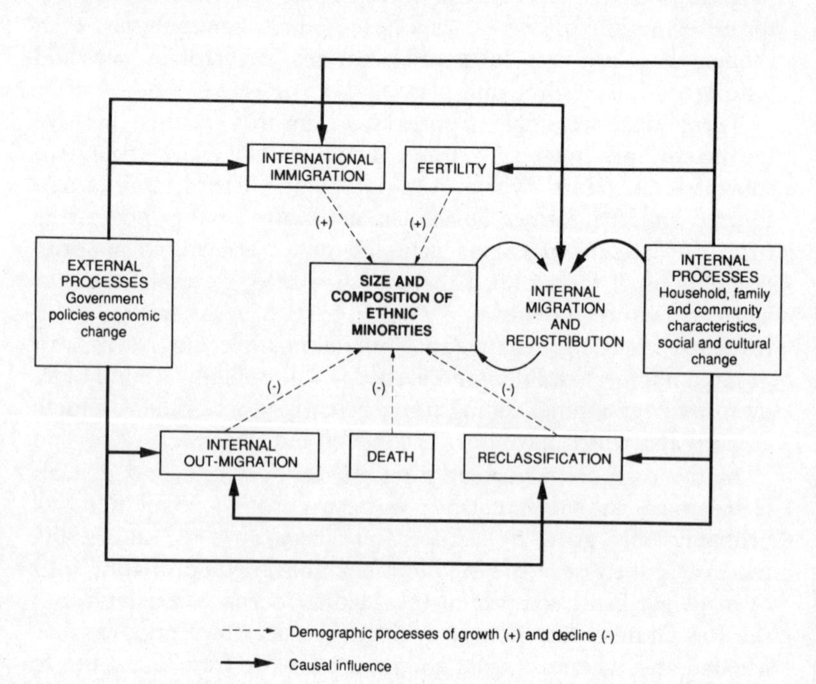


**Figure 17.1** Influences on, and processes of, demographic change in ethnic minority communities.

population will also show higher losses through mortality. However, the abundance of young people in most minority communities in western Europe is such that, despite reducing fertility levels, natural change will continue to be positive for some years to come. The other force 'threatening' the size of minority communities is highly variable from group to group and country to country: for reclassification depends on both the existence of a suitable political framework and the will of those involved to take action. Reclassification is the process leading to the greatest complexity with many mistakes between administrative and social reality. The external controls on reclassification, often in the form of naturalization, flow from government. The internal processes are those conditioning the self-identification of individuals, as in the 1991 British census. It is these processes that have the greater social validity, although the administrative classifications are permissible for a whole series of bureaucratic details affecting the daily lives of those involved.

The final demographic process shown in figure 17.1 is the vital geographical one of internal migration which will influence the location of communities and their degrees of dispersion or segregation. The conditioning influences here are, again, twofold with official policies being of particular significance in manipulating the housing market, but with the changing aspirations and demands of individuals from the minority communities being balancing factors.

The particular blend of demographic processes operating will vary from group to group. For example, for what are likely to be the growing numbers of highly skilled circulatory migrants it will be in- and out-migration that will control the sizes of minority groups in particular places at particular times. On the other hand, Rees and Ram (1987), in their ambitious predictive model for the future of Indian settlement in Bradford, believe that while immigration and fertility have been seen as the crucial factors in the past, internal redistribution and fertility will take the prime roles in the future.

Discussions of ethnic minority evolution have now moved a long way beyond 1960s views on assimilation and integration. Western European societies are now multicultural in character and assimilationist notions are implausible in even the medium-term perspective of the next 30 years. But social scientists need to be alert to the variety of adjustment mechanisms taking place at both the community and the individual scale and to the contexts of political, cultural, economic and social control within which these mechanisms work.

# 18

# Postscript: One Europe, Many Demographies?

Robert Woods

It is Sunday 20 September 1992. The French are voting on the future of the Maastricht Treaty. The lira and the pound have been devalued. Civil war rages in Bosnia and 'ethnic cleansing' continues unchecked. There have been riots against refugees and foreign workers in Germany. Nationalist terrorists operate in Ireland and Spain. Red Army soldiers are still based in East Germany, but not Hungary and Czechoslovakia. The cruise missiles have gone. Europe has no military dictators of the right or left, just various forms of parliamentary democracy. Average living standards are higher than they have ever been, but income distributions are still skewed and unemployment rates have been increasing; they are even more socially and regionally divisive. There are more qualifications, but fewer jobs.

At any moment in time it is difficult to distinguish the permanent from the transitory. Economic circumstances will change in a cyclical fashion that may be distorted by random crises. But Europe's regional economies are becoming more and more closely integrated. The core countries have core regions of economic activity which are parts of national and international networks. The south and selected parts of the periphery are experiencing perhaps temporary revival; the remainder, especially the peripheral industrial regions, face continued decline and further depopulation. There is no single currency or foreign policy, but English is increasingly taught as the second language. Attitudes to food and drink vary, but chips, beefburgers and pizzas are ubiquitous. Shops, houses, flats, towns, cars and clothes increasingly look alike. There is a European style, with national modifications, created in Paris, London or Milan, and monitored in Brussels. There is no state communism, but the churches are empty; belief is unfashionable, pessimism and cynicism rule in their place.

Of all these contradictory changes, the demographic may prove to be some of the most significant and certainly the most far-reaching. Although there are still relative differences in mortality, absolute levels are very low everywhere. Death comes in old age. Fertility is also low and where it varies it is the coincidence of timing that is most significant rather than differences in ideal family size. Cohabitation complements marriage while divorce becomes more common. Access to effective contraceptives has never been better. Fertility control is now at its most effective. Women can choose, their presence in the labour force has never been greater, but there have never been so many female headed single-parent households or so many widows as there are today. Average household size diminishes as the young and the old live on their own. The population is ageing everywhere in Europe.

There are still families and communities, Christmas and Easter are still celebrated, but television and sport have more unifying roles. Even the traditional scapegoats are present although they may have taken different faces: refugees, North Africans, Turks, those with AIDS, single parents, the homeless, the long-term unemployed, strangers. These are the new marginalized populations, the new outcasts to be kept at bay.

This book has mainly been concerned with what Fernand Braudel called *conjonctures*, although it has also dealt with certain important events, especially those associated with political changes and new legislation, but what of the *longue durée* and the *histoire démographique immobile*? Europeans are accustomed to thinking of themselves as inhabitants of the world's cultural heartland. They have grudgingly accepted their recent political and military subordination to the USA and the USSR, while the new CIS is in its turn adjusting to its new subordinate position. In demographic terms, there are now more Americans and Africans than Europeans. Figure 18.1 shows what is currently thought to have been the course and geographical components of world population growth in the very long term. Europe will almost certainly be the first world region to reach the relative security of zero population growth in a new age, post-industrial, post-modern society. The demographic upheavals of the last two hundred years will be set in their correct context: major, but quite short-run perturbations.

What are the principal characteristics of the new European demographic regime? First, there is unrestricted movement of European Community nationals, as well as those from the

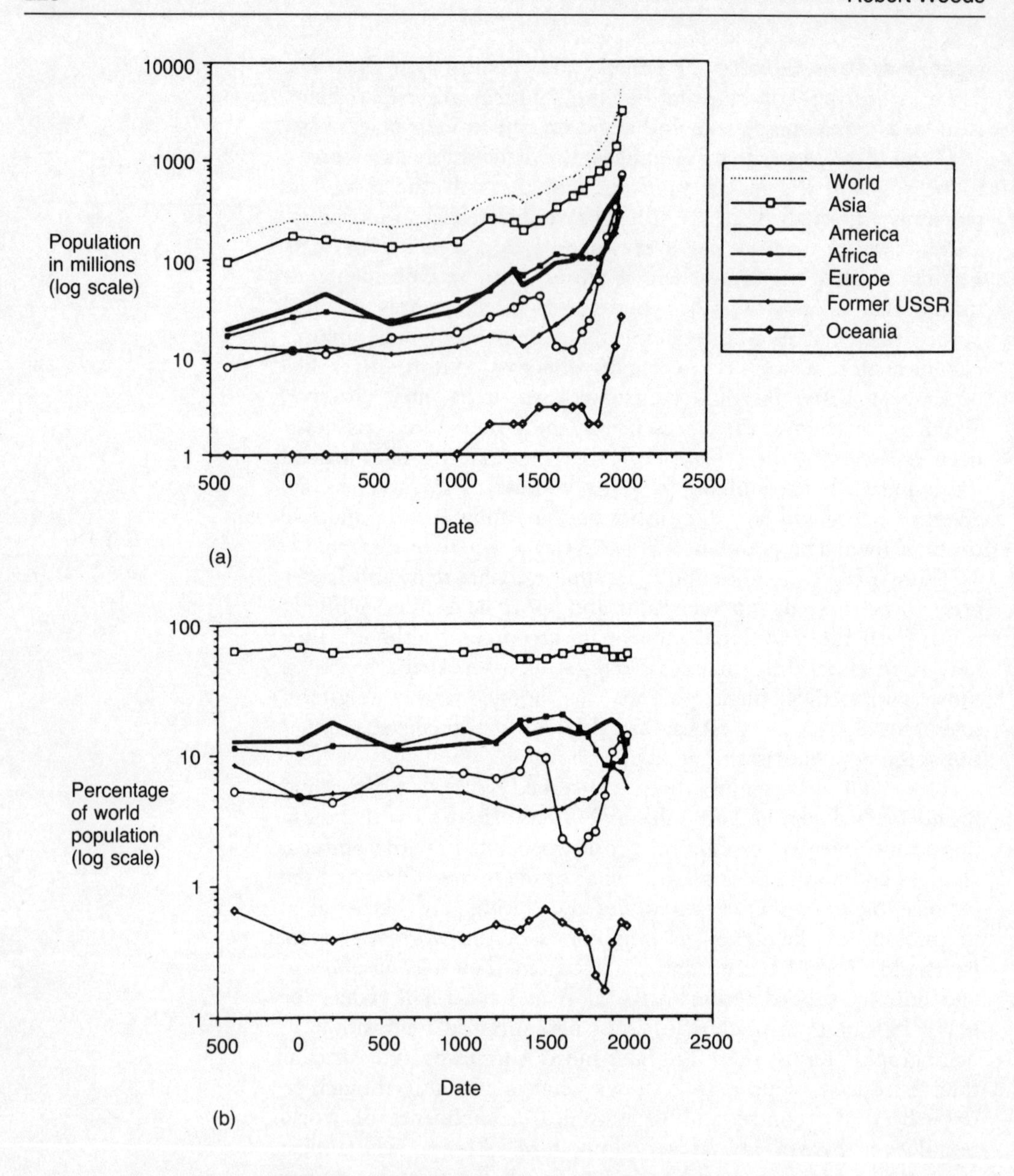


**Figure 18.1** (a) Trends in world population growth by major regions; (b) changes in the relative distribution of world population.

European Free Trade Area, within the Community. However, migration remains limited by the traditional constraints – employment opportunities, information, inertia, accommodation and so forth – but social segregation has if anything become more accentuated in the cities and suburbs. Enhanced mobility has helped to create separate social worlds. Second, only the old die. The survivorship curve with age is now almost rectangular;

without cancer and heart disease it would be even more so. The habits of diet and smoking, the inequities of housing, hazards at work and in the wider environment all help to support social differentials in health and morbidity which still have an important bearing on mortality differentials. Third, the new European marriage pattern is one dominated by cohabitation and divorce. Robert Malthus's marriage pattern, the preventive check that made north-west Europe so demographically distinct for probably a thousand years, has faded in the last half century. Fourth, most European women are, for most of their lives, infertile. The phenomenon of *Sex* has only been possible as a consequence. Pregnancies are largely planned to minimize dislocation costs and timed to coincide with improving economic circumstances. This regime is now firmly in place and, since it maximizes individual choice in a way that its predecessor could never have achieved, there is little reason to believe that its features will not remain in place for some time to come.

The French have voted yes. Political integration may follow the convergence to one European demographic regime, but at a safe distance and with hesitation.

# Bibliography

Adamo, F. (1986) The urban crisis in Italy. In G. Heinritz and E. Lichtenberger (eds) *The Take-Off of Suburbia and the Crisis of the Central City* (Wiesbaden: Steiner), pp. 207–27.

Alderson, M. and Ashwood, F. (1985) Projections of mortality rates for the elderly. *Population Trends* 42, pp. 22–9.

Als, G. (1990) La politique familiale au Luxembourg. In W. Dumon (ed.) *Family Policy in EEC Countries* (Luxembourg: Office for Official Publications of the European Community), pp. 253–76.

Angelini, E. and Magni, C. (1984) Il mercato del lavoro in alcune microaree italiane: aspetti metodologici e principali risultati. In M. Schenkel (ed.) *L'Offerta di Lavoro in Italia* (Venice: Marsillo), pp. 167–96.

Arena, G. (1982) Lavoratori stanieri in Italia e a Roma. *Bollettino, Societá Geografica Italiana* 11, pp. 57–93.

—— (ed.) (1990) *Geografia al Femminile* (Milan: Unicopli).

Atzema, O. and Buursink, J. (1985) The regional redistribution of Mediterraneans in the Netherlands. In P. E. White and G. A. van der Knaap (eds) *Contemporary Studies of Migration* (Norwich: GeoBooks), pp. 27–44.

Bagnasco, A. (1981) La questione dell economia informale. *Stato e Mercato* 1, pp. 173–96.

Bähr, J. and Gans, P. (1987) Development of the German and foreign population in the larger cities of the Federal Republic of Germany since 1970. In G. Glebe and J. O'Loughlin (eds) *Foreign Minorities in Continental European Cities* (Wiesbaden: Steiner), pp. 90–115.

Baines, D. (1991) *Emigration from Europe, 1815–1930* (London: Macmillan).

Barbier, J. C. (1989) La protection sociale de la famille dans les pays de la Communauté. *Revue Française des Affaires Sociales* (November), pp. 71–2.

—— (1990a) Comment comparer les politiques familiales en Europe: quelques problèmes de méthodes. *Revue Internationale de Sécurité Sociale* 3, pp. 342–57.

—— (1990b) Pour bien comparer les politiques familiales en Europe. *Revue Française des Affaires Sociales* 3 (July–September), pp. 153–71.

Barile, G. (ed.) (1984) *Lavoro Femminile, Sviluppo Tecnologico e Segregazione Occupazionale* (Milan: Angeli).

Barsotti, O. and Lecchini, L. (1989) L'immigration des pays du Tiers-Monde en Italie. *Revue Européenne des Migrations Internationales* 5, pp. 45–63.

Beale, C. L. (1982) *US Population: Where We Are, Where We're Going, Population Bulletin* 37 (2).

Becker, G. S. (1960) An economic analysis of fertility. In National Bureau for Economic Research, *Demographic and Economic Change in Developed Countries* (Princeton: Princeton University Press).

Bell, D. (1973) *The Coming of Post-Industrial Society* (New York: Basic Books).

—— (1986) La terza rivoluzione tecnologica e le possibili conseguenze socio-enonomiche. In G. Tamburini (ed.) *Occupazione e Tecnologie Avanzate* (Bologna: Il Mulino), pp. 47–73.

Beneria, L. (1990) Gender and the global economy. *Traballs de la Societat Catalana de Geografia* 23.

Benjamin, B. (1988) Demographic aspects of ageing. *Annals of Human Biology* 16, pp. 185–235.

Berent, J. (1983) *Family Size Preferences in Europe and the USA: Ultimate Expected Number of Children, World Fertility Survey, Comparative Studies* (Voorburg: International Statistical Institute).

Bergues, H. (1973) L'immigration de travailleurs africains noirs en France et particulièrement dans la région parisienne. *Population* 28, pp. 59–79.

Berry, B. J. L. (1976) *Urbanization and Counterurbanization* (Beverly Hills: Sage Publications).

Bertrand, J.-R. (1983) Les Galiciens en Suisse, 'aventure solitaire' ou migration collective? *Espace, Populations, Sociétés* 1 (2), pp. 39–47.

Bhat, A., Carr-Hill, R. and Ohri, S. (1988) *Britain's Black Population: A New Perspective* (Aldershot: Gower).

Bird, E. (1980) *Information Technology in the Office. The Impact on Women's Jobs* (Manchester: Equal Opportunity Commission).

Blanc, M. (1983) Le logement des travailleurs immigrés en France: après le taudis, le foyer et aujourd'hui le HLM. *Espace et Sociétés* 42, pp. 129–40.

Blanchet, D. (1987) Les effets démographiques de différentes mesures de politique familiale: une essai d'évaluation. *Population* 42, pp. 99–127.

Blayo, C. (1991) Les modes de prévention des naissances en Europe de l'Est. *Population* 46, pp. 527–46.

Boh, K. (ed.) (1989) *Changing Patterns of European Family Life: A Comparative Analysis of 14 European Countries* (London: Routledge).

Böhning, W. R. (1972) *The Migration of Workers in the United Kingdom and in the European Community* (London: Oxford University Press).

Bolzman, C., Fibbi, R. and Garcia, C. (1987) La deuxième génération d'immigrés en Suisse: catégorie ou acteur social?. *Revue Européenne des Migrations Internationales* 3, pp. 55–72.

Booth, H. (1985a) Which 'ethnic question? The development of questions identifying ethnic origin in official statistics. *Sociological Review* 33, pp. 154–74.

—— (1985b) *Second Generation Migrants in Western Europe: Demographic Data Sources and Needs.* Statistical Papers in Ethnic Relations No. 1, Centre for Research in Ethnic Relations, University of Warwick.

Boserup, E. (1982) *Il Lavoro delle Donne* (Turin: Rosenberg and Sellier).

Bourgeois-Pichat, J. (1981) Recent demographic change in Western Europe: an assessment. *Population and Development Review* 7, pp. 19–42.

Bouvier-Colle, M. H. and Jougla, E. (1989) Étude européenne de la répartition géographique des causes de décès 'évitables'. In *Géographie et Socioéconomie de la Santé* (Paris: CREDES).

Bradford, M. and Burdett, F. J. (1990) The geography of the changes in the consumption cleavage between state and private education in England. *Espace, Populations, Sociétés* 8 (1), pp. 33–48.

Brody, J. A. (1985) Prospects for an aging population. *Nature* 315, pp. 463–6.

Brouard, N. and Lopez, A. (1985) Cause of death in low mortality countries: a classification analysis. *International Population Conference, New Delhi, 1985*, vol. 2 (Liège: IUSSP), pp. 385–406.

Brunetta, G. and Rotondi, G. (1989) Différentiation régionale de la fécondité italienne depuis 1950. *Espace, Populations, Sociétés* 7 (2), pp. 189–200.

Bulusu, L. (1991) International migration in 1989. *Population Trends* 63, pp. 40–3.

Burnel, R. (1986) *Demographic Situation in the Community* (Brussels: Economic and Social Consultative Assembly).

—— (1990) *La Politique Familiale Globale, Conférence de l'Union Internationale des Organismes Familiaux, Moscou, 19–21 Octobre 1990* (Paris: Union Nationale des Associations Familiales).

Butcher, I. J. and Ogden, P. E. (1984) West Indians in France: migration and demographic change. In P. E. Ogden (ed.) *Migrants in Modern France: Four Studies*, Occasional Paper No. 23, Department of Geography and Earth Science, Queen Mary and Westfield College, University of London, pp. 43–66.

Büttner, T. and Lutz, W. (1990) Estimating fertility responses to policy measures in the German Democratic Republic. *Population and Development Review* 16, pp. 539–55.

Calot, G. and Deville, J. C. (1971) Nuptialité et fécondité selon le milieu socio-culturel. *Economie et Statistique* 27, pp. 3–42.

Campbell, A. (1974) Beyond the demographic transition. *Demography* 11, pp. 549–61.

Caselli, G. and Egidi, V. (1981) New studies in European mortality. *Demographic Studies* 5 (Strasbourg: Council of Europe).

Casper, W. and Hermann, S. (1991) The development of life expectancy in European countries. In *Socioeconomic Differential Mortality in Industrialized Societies*, vol. 7 (Paris: CICRED), pp. 215–26.

Castles, S. (1986) The guest-worker in Western Europe – an obituary. *International Migration Review* 20, pp. 761–78.

Castles, S., Booth, H. and Wallace, T. (1984) *Here for Good: Western Europe's New Ethnic Minorities* (London: Pluto Press).

Champion, A. G. (ed.) (1989a) *Counterurbanization: The Changing Pace and Nature of Population Concentration* (London: Edward Arnold).

—— (1989b) Counterurbanization in Britain. *Geographical Journal* 155, pp. 51–9.

—— (1991) Changes in the spatial distribution of the European population. *Proceedings of the Seminar on Present Demographic Trends and Lifestyles in Europe*, 18–20 September 1990 (Strasbourg: Council of Europe), pp. 355–88.

—— (1992) Urban and regional demographic trends in the Developed World. *Urban Studies* 29, pp. 461–82.

Champion, A. G. and Congdon, P. D. (1988) Recent population trends for Greater London. *Population Trends* 53, pp. 11–17.

Cheshire, P. C. and Hay, D. G. (1989) *Urban Problems in Western Europe: An Economic Analysis* (London: Unwin Hyman).

Chesnais, J.-C. (1985) Les conditions d'efficacité d'une politique nataliste: examen théorique et examples historiques. *International Population Conference, Florence, 1985*, vol. 3 (Liège: IUSSP), pp. 413–25.

—— (1986) *La Transition Démographique* (Paris: Presses Universitaires de France).

—— (1990) Demographic change in Europe and its social and economic consequences. Paper presented at the Conference on the United States and Europe in the 1990s: Trade, Finance, Defence, Politics, Demographics and Social Policy, Washington, DC, 5–8 March 1990.

Clarke, J. I. (1987) Ageing in Europe: introductory remarks. *Espace, Populations, Sociétés* 5 (1), pp. 23–8.

Clason, C. F. (1986) One-parent families in the Netherlands. In F. Deven and R. L. Cliquet (eds) *One-Parent Families in Europe: Trends, Experiences, Implications* (The Hague/Brussels: NIDI/CBGS), pp. 195–208.

Coale, A. J. and Watkins, S. C. (eds) (1986) *The Decline of Fertility in Europe* (Princeton: Princeton University Press).

Cochrane, S. G. and Vining, D. R. (1988) Recent trends in migration between core and peripheral regions in developed and advanced developing countries. *International Regional Science Review* 11, pp. 215–44.

Cohen, R. (1987) *The New Helots. Migration in the International Division of Labour* (Aldershot: Gower).

Coleman, D. A. (1985) Ethnic intermarriage in Great Britain. *Population Trends* 40, pp. 4–10.

—— (1991) European demographic systems of the future: convergence or diversity? Paper presented at the conference on *Human Resources in Europe at the Dawn of the 21st Century*, Luxembourg, 27–9

November 1991.
—— (1992) Does Europe need immigrants? Population and work force projections. *International Migration Review* 26, pp. 413–61.
Coleman, D. A. and Salt, J. (1992) *The British Population: Patterns, Trends and Processes* (Oxford: Oxford University Press).
Collicelli, C. and di Cori, S. (1986) L'immigrazione staniera in Italia nel contesto delle problematiche migrarorie internazionali. *Studi Emigrazione* 82–3, pp. 429–36.
Commaille, J. (1990) Family policy and population policy. In W. Dumon (ed.) *Family Policy in EEC Countries* (Luxembourg: Office for Official Publications of the European Community), pp. 59–77.
Commission des Communautés Européenes (1989) *Communication de la Commission sur les Politiques Familiales*, COM (89) 363 final, Documents FR 05 (Luxembourg: Office des Publications Officielles des Communautés Européennes).
Commission of the European Community (1981) *Le Marché du Travail Européen* (Brussels).
—— (1990) *Family Policy in EC Countries*. Report prepared for the Commission of the European Community, Director General for Employment, Social Affairs and Education (Luxembourg: Office for Official Publications of the European Community).
Conseil Economique et Social (1991) *La Politique Familiale Française*. Report presented by M. Hubert Brin at the meetings of 24–5 September 1991 (Paris: Journal Officiel de la République Française).
Coopmans, M., Harrop, A. and Hermans-Huiskes, M. (1988) *The Social and Economic Situation of Older Women in Europe* (Brussels: CEC).
Council of Europe (1985) *Changing Age Structure of the Population and Future Policy*, Population Studies 18 (Strasbourg: Council of Europe).
—— (1990a) *Household Structures in Europe*, Population Studies 22 (Strasbourg: Council of Europe).
—— (1990b) *Recent Demographic Developments in the Member States of the Council of Europe, 1989* (Strasbourg: Council of Europe).
Courgeau, D. (1985) Interaction between spatial mobility, family and career life-cycle: a French survey. *European Sociological Review* 1 (2), pp. 139–62.
Court, Y. (1986) Denmark. In A. Findlay and P. E. White (eds) *West European Population Change* (London: Croom Helm), pp. 81–101.
Curtis, S. E. and Ogden, P. E. (1986) Bangladeshis in London: a challenge to welfare. *Revue Européenne des Migrations Internationales* 2, pp. 135–50.
Dahlstrom, E. (1989) Theories and ideologies of family functions, gender relations and human reproduction. In K. Boh *et al.* (eds) *Changing Patterns of European Family Life: A Comparative Analysis of 14 European Countries* (London: Routledge), pp. 31–51.
Davis, K., Bernstam, M. S. and Ricardo-Campbell, R. (eds) (1987) *Below-Replacement Fertility in Industrial Societies: Causes, Consequences, Policies* (Cambridge: Cambridge University Press).
Decroly, J. M. and Vanlaer, J. (1991) *Atlas de la Population Européenne* (Brussels: Editions de l'Université de Bruxelles).

Del Campo, S. (1990) Current family policy in Spain. In W. Dumon (ed.) *Family Policy in EEC Countries* (Luxembourg: Office for Official Publications for the European Community), pp. 335–49.

Desplanques, G. (1984) L'inégalité sociale devant la mort. *Economie et Statistique* 179, pp. 26–49.

Diáz Alvarez, J. R. (1989) Presencia y significatión del emigrado español en el Reino de Bélgica. *Estudios Geograficos* 194, pp. 35–64.

Dielman, F. M., de Klerk, L. and Clark, W. A. V. (1991) School segregation, forced integration and districting, or free choice?. Paper presented to the International Seminar on Population Geography, Soesterberg, The Netherlands.

Donati, P. (1990) Family and population policy in Italy. In W. Dumon (ed.) *Family Policy in EEC Countries* (Luxembourg: Office for Official Publications of the European Community), pp. 207–52.

Dooghe, G. (1991) *The Ageing of the Population of Europe: Socio-Economic Characteristics of the Population*, Centrum voor Bevolkings en Gezinsstudien, Ministerie Vlaamse Gemeenschap (Brussels).

Dumon, W. (ed.) (1990) *Family Policy in EEC Countries* (Luxembourg: Office for Official Publications of the European Community).

Dupâquier, J. (ed.) (1988) *Histoire de la Population Française*, vol. 4 (Paris: Presses Universitaires de France).

Easterlin, R. (1976) The conflict between aspirations and resources. *Population and Development Review*, 2, pp. 417–25.

—— (1980) *Birth and Fortune* (New York: Basic Books).

Ekert-Jaffé, O. (1986) Effets et limites des aides financières aux familles: une expérience et un modèle. *Population* 41, pp. 327–48.

—— (1988) Les politiques familiales en Europe: objectifs et entités culturelles. In Ministère de la Solidarité, de la Santé et de la Projection Sociale, *Études Statistiques* 4 (July–August), *Dossier: Famille et Politiques Familiales* (Paris), pp. 47–58.

Entzinger, H. B. (1985) The Netherlands. In T. Hammar (ed.) *European Immigration. A Comparative Study* (Cambridge: Cambridge University Press), pp. 50–88.

Epstein, T. S. (1986) *Women, Work and Family in Britain and Germany* (London: Croom Helm).

Equal Opportunity Commission (1980) *Equality for Women, Assessment, Problem, Perspectives: A European Project* (Manchester: Equal Opportunity Commission).

Ermisch, J. (1989) Divorce: economic antecedents and aftermath. In H. Joshi (ed.) *The Changing Population of Britain* (Oxford: Basil Blackwell), pp. 42–55.

—— (1990a) European women's employment and fertility again. *Journal of Population Economics* 3, pp. 3–18.

—— (1990b) *Fewer Babies: Longer Lives* (York: Joseph Rowntree Foundation).

EUROSTAT (1986) *Education and Training, 1985* (Brussels: EUROSTAT).

—— (1989a) *Employment and Unemployment* (Brussels: EUROSTAT).

—— (1989b) *Labour Force Survey. Results, 1987* (Brussels: EUROSTAT).

—— (1989c) *Regions. Statistical Yearbook, 1988* (Brussels: EUROSTAT).

—— (1990) *Demographic Statistics*, Theme 3, Series C (Brussels: EUROSTAT).

Fagnani, J. (1980) City size and mother's labour force participation. *Tijdschrift voor Economische en Sociale Geografie* 3, pp. 182–3.

Federici, N. (1984) *Procreazione, Famiglia, Lavoro della Donna* (Turin: Loeseher).

Festy, P. (1979) *La Fécondité dans les Pays Occidentaux de 1870 à 1970*, INED 85 (Paris: Presses Universitaires de France).

—— (1983) Le mouvement quadriennal des mariages en Grèce. *Population* 38, pp. 409–13.

—— (1985) *Le Divorce, la Séparation Judiciaire et le Remarriage. Evolution Récente dans les États Membres du Conseil de l'Europe*, Demographic Studies 17 (Strasbourg: Council of Europe).

—— (1986) Fécondité et politiques démographiques en Europe de l'Est. *Politiques de Population, Études et Documents* 2, pp. 7–51.

Fielding, A. J. (1982) Counterurbanization in Western Europe. *Progress in Planning* 17, pp. 1–52.

—— (1986) Counterurbanization in Western Europe. In A. Findlay and P. E. White (eds) *West European Population Change* (London: Croom Helm), pp. 35–49.

Fincher, R. (1989) Class and gender relations in the local labour market and the local state. In J. Wolch and M. Dear (eds) *The Power of Geography* (London: Unwin Hyman).

Findlay, A. and Garrick, L. (1990) Scottish emigration in the 1980s: a migration channels approach to the study of skilled international migration. *Transactions of the Institute of British Geographers* 15, pp. 177–92.

Findlay, A. and White, P. E. (eds) (1986) *West European Population Change* (London: Croom Helm).

Fischer, M. and Niskamp, P. (1987) Cross-national comparison of regional labour markets in 15 countries. In M. Fischer and P. Niskamp (eds) *Regional Labour Markets. Analytical Contributions and Cross-National Comparisons* (Amsterdam: North-Holland).

Förderzentrum Jugend Schreibt (1980) *Täglich eine Reise von der Türkei nach Deutschland: Texte der Zweiten Türkischen Generation der Bundesrepublik* (Fischerhude: Atelier in Bauerhaus).

Fox, A. J. (ed.) (1989) *Health Inequalities in European Countries* (Aldershot: Gower).

Frejka, T. (1983) Induced abortion and fertility: a quarter century of experience in Eastern Europe. *Population and Development Review* 9, pp. 494–520.

Frey, M. (1990) Ausländerpolitiken in Europa. In C. Höhn and D. B. Rein (eds) *Ausländer in der Bundesrepublik Deutschland*, Schriftenreihe des Bundesinstituts für Bevölkerungsforschung 20 (Boppard: Boldt-Verlag), pp. 121–47.

Frey, M. and Lubinski, V. (1987) *Probleme infolge hoher Ausländerkonzentration in ausgewählten europäischen Staaten*,

Materialien zur Bevölkerungswissenschaft, Sonderheft 8 (Wiesbaden: Bundesinstitut für Bevölkerungsforschung).

Fries, J. F. (1980) Aging, natural death and the compression of morbidity. *New England Journal of Medicine* 303, pp. 130–6.

—— (1989) The compression of morbidity: near or far. *Milbank Memorial Fund Quarterly* 67, pp. 208–32.

Gambier, D. (1980) Marché du travail et espace: un point de vue théorique. *L'Espace Géographique* 9 (1), pp. 1–13.

Gambino, F. (1984) L'Italie, pays d'immigration: rapports sociaux et formes juridiques. *Peuples Mediterranéens, Mediterranean Peoples* 27–8, pp. 173–85.

Gans, P. (1990) Changes in the structure of the foreign population of West Germany since 1980. *Migration* 7, pp. 25–49.

Garcia Ballesteros, A. (1979) Ocupación y paro en Madrid y en la región Castellano-Manchega. *Estudios Geográficos* (1979), pp. 156–7.

—— (1991) Descenso de la fecundidad y paro en España (1972–1990). In *Homenaje a Angel Cabo* (Salamanca: Universidad Salamanca).

Garcia Ballesteros, A. and Crespo Valero, M. J. (1988) Chômage et fécondité post-transitionelle en Espagne. Presented at a seminar on *La Transition Démographique dans le Pays Mediterrannées* (Nice).

Garcia Ballesteros, A. *et al.* (1985) Activité et chômage en Espagne. Contrastes dans l'espace et le temps (1955–1984). *Espace, Populations, Sociétés* 3 (2), pp. 357–74.

Gershuny, J. (1980) *Technical Innovation and Women's Work in the EEC. A Medium-Term Perspective* (Brighton: University of Sussex).

Giese, E. (1978) Räumliche Diffusion ausländischer Arbeitsnehmer in der Bundesrepublik Deutschland, 1960–76. *Die Erde* 109, pp. 92–110.

Gillion, C. (1991) Ageing populations: spreading the costs. *Journal of European Social Policy* 1 (2), pp. 107–28.

Gillis, J. R., Tilly, L. A. and Levine, D. (eds) (1992) *The European Experience of Declining Fertility: A Quiet Revolution, 1850–1970* (Oxford: Basil Blackwell).

Girard, A. (1977) Opinion publique, immigration et immigrés. *Ethnologie Française* 7, pp. 219–28.

Glaude, M. (1991) L'originalité du système du quotient familial. *Economie et Statistique* 248 (November), pp. 51–67.

Glebe, G. (1986) Segregation and intra-urban mobility of a high-status ethnic group: the case of the Japanese in Düsseldorf. *Ethnic and Racial Studies* 9, pp. 461–83.

—— (1990) Segregation and migration of the second generation of guestworker minorities in Düsseldorf. *Espace, Populations, Sociétés* 8 (2), pp. 257–8.

Glebe, G. and O'Loughlin, J. (eds) (1987) *Foreign Minorities in Continental European Cities* (Wiesbaden: Steiner).

Golini, A. (1987) Familie et ménage dans l'Italie récente. *Population* 42, pp. 699–714.

Golini, A. et al. (1991) *Famiglia, Figli e Società in Europa. Crisi della Natalità e Politiche per la Popolazione* (Turin: Edizioni della Fondazione Giovanni Agnelli).

Gonvers, J.-P. et al. (1981) La deuxième génération d'Espagnols en Suisse: résultats et interrogations d'une enquête-participation. *Schweizerische Zeitschrift für Soziologie* 7, pp. 279–92.

Gozálvez Pérez, V. (1990) El reciente incremento de la población extranjera en España y su incidencia laboral. *Investigaciones Geográficas* 8, pp. 7–36.

Grasland, C. (1990) Systèmes démographiques et systèmes supranationaux: la fécondité européenne de 1952 à 1989. *European Journal of Population* 6, pp. 163–91.

Guillon, M. (1988) Etrangers et immigrés dans la population de la France. In C. Wihtol de Wenden (ed.) *La Citoyenneté* (Paris: Edilig), pp. 17–56.

Haavio-Mannila, E. (1989) Gender segregation in paid and unpaid work. In K. Boh (ed.) *Changing Patterns of European Family Life: A Comparative Analysis of 14 European Countries* (London: Routledge), pp. 123–40.

Hagendoorn, L. and Hraba, J. (1989) Foreign, different, deviant, seclusive and working class: anchors to an ethnic hierarchy in the Netherlands. *Ethnic and Racial Studies* 12, pp. 441–68.

Hall, P. and Hay, D. (1980) *Growth Centres in the European Urban System* (London: Heinemann).

Hall, R. (1986) Household trends within Western Europe, 1970–1980. In A. Findlay and P. E. White (eds) *West European Population Change* (London: Croom Helm), pp. 18–34.

—— (1988) Recent patterns and trends in European households at national and regional scales. *Espace, Populations, Sociétés* 6 (1), pp. 13–32.

Hammar, T. (ed.) (1985) *European Immigration. A Comparative Study* (Cambridge: Cambridge University Press).

Hargreaves, A. (1989) Resistance and identity in *Beur* narratives. *Modern Fiction Studies* 35, pp. 87–102.

Hart, N. (1986) Inequalities in health: the individual versus the environment. *Journal of the Royal Statistical Society A* 149 (3), pp. 228–46.

Haskey, J. (1983) Marital status before marriage and age of marriage: their inflence on the chance of divorce. *Population Trends* 32, pp. 4–14.

—— (1986) One-parent families in Great Britain. *Population Trends* 45, pp. 5–13.

—— (1988a) Trends in marriage and divorce and cohort analyses of the proportions of marriages ending in divorce. *Population Trends* 54, pp. 29–31.

—— (1988b) The ethnic minority populations of Great Britain: their size and characteristics. *Population Trends* 54, pp. 29–31.

—— (1989a) Current prospects for the proportion of marriages ending in divorce. *Population Trends* 55, pp. 34–7.

—— (1989b) One-parent families and their children in Great Britain: numbers and characteristics. *Population Trends* 55, pp. 27–33.

—— (1989c) Families and households of the ethnic minority and white

populations of Great Britain. *Population Trends* 57, pp. 8–19.
—— (1991) Ethnic minority populations resident in private households – estimates by county and metropolitan district of England and Wales. *Population Trends* 63, pp. 22–35.
Haskey, J. and Kiernan, K. (1989) Cohabitation in Great Britain: characteristics and estimated numbers of cohabiting partners. *Population Trends* 58, pp. 23–32.
Hecht, J. (1975) *Législations Affectant Directement ou Indirectement la Fécondité en Europe* (Strasbourg: Council of Europe).
—— (1991) Il caso francese. In A. Golini et al. (eds) *Famiglia, Figli e Società. Crisi della Natalità e Politiche per la Popolazione* (Turin: Edizioni della Fondazione Giovanni Agnelli), pp. 137–67.
Heilig, G., Büttner, T. and Lutz, W. (1990) *Germany's Population: Turbulent Past, Uncertain Future, Population Bulletin* 25 (4).
Heisler, B. S. (1986) Immigrant settlement and the structure of emergent immigrant communities in Western Europe. *Annals of the American Academy of Political and Social Science* 485, pp. 76–86.
Hendricks, J. and Hendricks, C. D. (1986) *Aging in Mass Society: Myths and Realities* (Boston: Little, Brown & Co.).
Henripin, J. and Lapierre-Adamcyk, E. (1974) *La Fin de la Revanche des Berceaux: Qu'en Pensent les Québécoises?* (Montréal: Presses de l'Université de Montréal).
Henwood, M., Rimmer, L. and Wicks, M. (1987) *Inside the Family: Changing Roles of Men and Women.* Occasional Paper 6 (London: Family Policy Studies Centre).
Herdegen, G. (1989) Aussiedler in der Bundesrepublik Deutschland: Einstellen und aktuelle Ansichten der Bundesbürger. *Informationen zur Raumentwicklung* 5, pp. 331–41.
Hoem, J. (1990) Social policy and recent fertility change in Sweden. *Population and Development Review* 16, pp. 735–48.
Hoffmann-Nowotny, H.-J. and Fux, B. (1991) Present demographic trends in Europe. In *Proceedings of the Seminar on Present Demographic Trends and Lifestyles in Europe*, 18–20 September 1990 (Strasbourg: Council of Europe), pp. 31–97.
Holland, W. W. (ed.) (1988) *European Community Atlas of 'Avoidable Death'* (Oxford: Oxford University Press).
Hopflinger, F. (1985) Changing marriage behaviour: some European comparisons. *Genus* 41 (3–4), pp. 41–64.
—— (1991) The future of household and family structures in Europe. *Proceedings of the Seminar on Present Demographic Trends and Lifestyles in Europe*, Strasbourg, 18–20 September 1990 (Strasbourg: Council of Europe).
Hudson, R., Rhind, D. and Mounsey, H. (1984) *An Atlas of European Affairs* (London: Methuen).
Hulko, J. (1990) Family policy in Western Europe. *Yearbook of Population Research in Finland* 28, pp. 5–27.
Humphreys, M. (1988) Voluntary childlessness: the problem of long-range prediction. In H. Moors and J. Schoorl (eds) *Lifestyles, Contraception and Parenthood* (The Hague: NIDI/CBGS).

Huss, M.-M. (1988) The popular ideology of the child in wartime France: the evidence of the picture postcard. In R. Wall and J. M. Winter (eds) *The Upheaval of War: Family, Work and Welfare in Europe, 1914–1918* (Cambridge: Cambridge University Press).

—— (1990) Pronatalism in the inter-war period in France. *Journal of Contemporary History* (January), pp. 39–68.

ILO (1986) *Economically Active Population: Estimates and Projections, 1950–2025* (Geneva: International Labour Office).

—— (1988) *Yearbook of Labour Statistics* (Geneva: International Labour Office).

INED (1982) *Natalité et Politiques de Population en France et en Europe de l'Est, Travaux et Documents* 98 (Paris: INED-PUF).

International Migration Review (1992) *The New Europe and International Migration* 26 (2).

Izquierdo Escribano, A. (1990) *Immigration en Espagne et Premiers Résultats du Programme de Regularisation*. Unpublished report MAS/WP2(90)3, Organisation for Economic Co-operation and Development (Paris: OECD).

Jallinoja, R. (1989) Women between the family and employment. In K. Boh (ed.) *Changing Patterns of European Family Life: A Comparative Analysis of 14 European Countries* (London: Routledge), pp. 95–122.

Jenson, J., Hagen, E. and Reddy, C. (ed.) (1988) *Feminization of the Labour Force* (Cambridge: Polity Press).

Johanet, G. (1982) La nouvelle politique familiale. *Droit Social* 6 (June).

Jones, P. (1991) The French census 1990: the southward drift continues. *Geography* 76, pp. 358–61.

Jones, P. N. (1990a) Recent ethnic German migration from Eastern Europe to the Federal Republic. *Geography* 75, pp. 249–52.

—— (1990b) West Germany's declining guestworker population: spatial changes and economic trends in the 1980s. *Regional Studies* 24, pp. 223–33.

Jonker, J. M. L. (1990) Family policy and population policy in the Netherlands. In W. Dumon (ed.) *Family Policy in EEC Countries* (Luxembourg: Office for Official Publications of the European Community), pp. 277–314.

Joshi, H. (1989a) The changing form of women's economic dependency. In H. Joshi (ed.) *The Changing Population of Britain* (Oxford: Basil Blackwell), pp. 157–76.

—— (ed.) (1989b) *The Changing Population of Britain* (Oxford: Basil Blackwell).

Kane, T. T. (1986) The fertility and assimilation of guestworker populations in the Federal Republic of Germany, 1961–81. *Zeitschrift für Bevölkerungswissenschaft* 12, pp. 99–131.

Keilman, N. (1987) Recent trends in family and household composition in Europe. *European Journal of Population* 3, pp. 297–325.

—— (1988) Dynamic household models. In N. Kielman, A. Kuijsten and A. Vossen (eds) *Modelling Household Formation and Dissolution* (Oxford: Clarendon Press), pp. 123–38.

Keilman, N., Kuijsten, A. and Vossen, A. (eds) (1988) *Modelling*

*Household Formation and Dissolution* (Oxford: Clarendon Press).
Keyfitz, N. and Flieger, W. (1990) *World Population Growth and Aging* (Chicago: Chicago University Press).
Kiernan, K. (1986) Leaving home: living arrangements of young people in six west European countries. *European Journal of Population* 2, pp. 177–84.
—— (1989) Who remains childless?. *Journal of Biosocial Science* 21, pp. 387–98.
—— (1992) The impact of family disruption in childhood on transitions made in young adult life. *Population Studies* 46, pp. 213–34.
King, R., Shuttleworth, I. and Strachan, A. (1989) The Irish in Coventry: the social geography of a relict community. *Irish Geography* 22, pp. 64–78.
Klinger, A. (1991) Les politiques familiales en Europe de l'Est. *Population* 46, pp. 511–26.
Knodel, J. E. (1974) *The Decline of Fertility in Germany, 1871–1939* (Princeton: Princeton University Press).
Knodel, J. and van de Walle, E. (1979) Lessons from the past: policy implications of historical fertility studies. *Population and Development Review* 5, pp. 217–45.
Koch, L. (1989) Impact of the reversal of the migration situation on the social structure of certain countries. The case of Italy. *International Migration* 27, pp. 191–201.
Kontuly, T. and Vogelsang, R. (1989) Federal Republic of Germany: the intensification of the migration turnaround. In A. G. Champion (ed.) *Counterurbanization* (London: Edward Arnold), pp. 141–61.
Kosinski, L. (1970) *The Population of Europe* (London: Longman).
Kunst, A. *et al.* (1988) Medical care and regional mortality differences within the countries of the European Community. *European Journal of Population*, 4, pp. 223–45.
Kunzmann, K. R. and Wegener, M. (1991) The pattern of urbanization in Western Europe, 1960–1990. *Berichte aus dem Institut für Raumplanung* 28 (Dortmund: Institut für Raumplanung Universität Dortmund).
Laroque, M. (1987) Les aspects familiaux de la politique sociale de la Communauté Européenne. *Revue Française des Affaires Sociales* 4, pp. 7–18.
—— (1988) Les politiques familiales en Europe. Report to the Council of Europe (December).
Laslett, P. (1985) Societal development and ageing. In R. H. Binstock and E. Shanas (ed.) *Handbook of Aging and the Social Sciences* (New York: Van Nostrand Reinhold), pp. 199–230.
Lazaro Araujo, L. and Molina Ibañez, M. (1986) *El Espacio de la Communidad Económica Europea. La Política Regional* (Madrid: Triviam).
Lebon, A. (1986) Les travailleurs étrangers en Europe. (Combien sont-ils? Qui sont-ils? Où travaillent-ils?). *Revue Européenne des Migrations Internationales* 2, pp. 169–84.
—— (1990) Ressortissants communautaires et étrangers des pays tiers

dans l'Europe des Douze. *Revue Européenne des Migrations Internationales* 6.

Ledent, J. and Courgeau, D. (1982) *Migration and Settlement: 15. France, Research Report RR-82-28* (Laxenburg: International Institute for Applied Systems Analysis).

Lefaucheur, N. (1986) How the one-parent families appeared in France. In F. Deven and R. L. Cliquet (eds) *One-Parent Families in Europe: Trends, Experiences, Implications* (The Hague/Brussels: NIDI/CBGS), pp. 73–81.

Leibenstein, H. (1974) An interpretation of the economic theory of fertility: promising path or blind alley? *Journal of Economic Literature* 12, pp. 457–79.

Leitner, H. (1990) Informal work on the streets of Vienna: the foreign newspaper vendors, *Espace, Population, Sociétés* 8 (2), pp. 221–9.

Leridon, H. and Villeneuve-Gokalp, C. (1988) Les nouveaux couples: nombre, caractéristiques et attitudes. *Population* 43, pp. 331–74.

Leridon, H. *et al.* (1987) *La Seconde Révolution Contraceptive: La Régulation des Naissances en France de 1950 à 1985, INED, Travaux et Documents*, 117 (Paris: INED-PUF).

Lesthaeghe, R. (1983) A century of demographic and cultural change in Western Europe: an exploration of underlying dimensions. *Population and Development Review* 9, pp. 411–35.

Lichtenberger, H. (1984) *Gastarbeiter: Leben in zwei Gesellschaften* (Vienna: Hermann Böhlau).

Lloyd. C. B. and Ross, J. A. (1987) Methods for measuring the fertility impact of family planning programs: the experience of the last decade. Population Council, Research Division Working Papers 7 (New York: Population Council).

Lutz, W. (ed.) (1990) *Future Demographic Trends in Europe and North America* (London: Academic Press).

Manacorda, P. M. and Piva, P. (1985) *Terminale Donna* (Rome: Edizioni Lavoro).

Manfrass, K. (1983) Ausländerpolitik in Frankreich seit 1981. *ZAR* 4, pp. 163–72.

—— (1992) Europe: South-North or East-West migration?. *International Migration Review* 26, pp. 388–400.

Martin, J. and Roberts, C. (1984) *Women and Employment: A Lifetime Perspective* (London: Department of Employment).

Martin, P. L., Hönekopp, E. and Ullmann, H. (1990) Europe 1992: effects on labour migration. *International Migration Review* 24, pp. 591–603.

Martin, R. L. (1982) Britain's slump: the regional anatomy of job loss. *Area* 14, pp. 101–7.

Mathews, G. (1989) *Politiques Natalistes Européennes et Politique Familiale Canadienne, Institut National de la Recherche Scientifique, Études et Documents*, 59 (Montréal: INRS).

McIntosh, C. A. (1983) *Population Policy in Western Europe: Responses to Low Fertility in France, Sweden and West Germany* (New York: M. E. Sharpe).

Meyze, C. and Rose, L. (1983) Les 'Bana-Bana', esclaves de nos trottoirs. *Hommes et Migrations* 34 (1051), pp. 25–31.

Mingione, E. (1985) Marginale e povero: il nuovo immigrato in Italia. *Politica ed Economia* 6, pp. 61–4.

Ministère de la Solidarité, de la Santé et de la Protection Sociale (1988) *Études Statistiques*, 4 (July–August), *Dossier: Familles et Politiques Familiale* (Paris), pp. 1–58.

Monnier, A. (1989) Bilan de la politique familiale en République Démocratique Allemande: un réexamen. *Population* 44, pp. 379–94.

—— (1990) *La Population de la France. Mutations et Perspectives* (Paris: Messidor).

—— (1991) L'Europe de l'Est différente et diverse. *Population* 46, pp. 443–62.

Montagné-Villette, S. (ed.) (1991) *Espaces et Travails Clandestins* (Paris: Masson).

Moors, H. and van Nimwegen, N. (1990) *Social and Demographic Effects of Changing Structures on Children and Young People* (Strasbourg: Council of Europe).

Morrison, P. S. (1990) Segmentation theory applied to local, regional and spatial labour markets. *Progress in Human Geography* 4, pp. 210–25.

Muñoz-Perez, F. (1986) Changements récents de la fécondité en Europe Occidentale et nouveau traits de la formation des familles. *Population* 41, pp. 447–62.

—— (1987) Le déclin de la fécondité dans le sud de l'Europe. *Population* 42, pp. 911–42.

Muñoz-Perez, F. and Izquierdo Escribano, A. (1989) L'Espagne, pays d'immigration. *Population* 44, pp. 257–89.

Muñoz-Perez, F. and Tribalat, M. (1984) Mariages d'étrangers et mariages mixtes en France: évolution depuis la première guerre. *Population* 39, pp. 427–62.

Murphy, M. (1992) Economic models of fertility in post-war Britain – a conceptual and statistical re-interpretation. *Population Studies* 46, pp. 235–57.

Murphy, M., Sullivan, O. and Brown, A. (1988) Sources of data for modelling household change with special reference to the OPCS 1 per cent Longitudinal Study. In N. Kielman, A. Kuijsten and A. Vossen (eds) *Modelling Household Formation and Dissolution* (Oxford: Clarendon Press), pp. 56–66.

Murray, J. (1989) Migration and European society: a view from the Council of Europe. *International Review of Comparative Public Policy* 1, pp. 179–88.

Muxel, A. (1988) Les attitudes socio-politiques des jeunes issues de l'immigration maghrébine en région parisienne. *Revue Française de Science Politique* 38, pp. 925–40.

Nasman, E. (1991) Il caso svedese. In A. Golini *et al.* (eds) *Famiglia, Figli e Società. Crisi della Natalità e Politiche per la Popolazione* (Turin: Edizione della Fondazione Giovanni Agnelli), pp. 169–208.

Nebe, J. M. (1988) Residential segregation of ethnic groups in West German cities. *Cities* 5, pp. 235–44.

Ní Bhrolcháin, M. (1990) The ethnicity question for the 1991 census: background and issues. *Ethnic and Racial Studies* 13, pp. 542–67.

Nijkamp, P., Pacolet, J. and Vollering, A. (eds) (1990) *Services for the Elderly in Europe* (Brussels: CEC).

Nobile, A. (1990) Recent trends in infant mortality in developed countries. *Genus* 46, pp. 79–107.

Noin, D. (1983) *La Transition Démographique dans le Monde* (Paris: Presses Universitaires de France).

—— (1988) *Géographie de la Population* (Paris: Masson).

—— (1991) *Atlas de la Population Mondiale* (Paris: Reclus).

—— (1992) *La Population de la France* (Paris: Masson).

Noin, D. and Warnes, A. M. (eds) (1987) *Elderly People and Ageing. Espace, Populations, Sociétés* 5 (1), special issue.

Norro, P. (1990) Accueil et répartition des candidats-réfugiés politiques en Belgique. *Espace, Populations, Sociétés* 8 (2), pp. 191–205.

Notestein, F. W., Kirk, D. and Taeuer, I. B. (1944) *The Future Population of Europe and the Soviet Union: Population Projections, 1940–1970* (London: Allen and Unwin, for the League of Nations).

Nuss, S., Denti, E. and Viry, D. (eds) (1989) Women in the world of work: statistical analysis and projections to the year 2000. *Women, Work and Development* 18 (Geneva: International Labour Office).

O'Higgins, K. (1990) Report on family/population policy in Ireland. In W. Dumon (ed.) *Family Policy in EEC Countries* (Luxembourg: Office for Official Publications of the European Community), pp. 171–206.

O'Loughlin, J. (1980) Distribution and migration of foreigners in German cities. *Geographical Review* 70, pp. 253–75.

OECD (1985) *The Integration of Women into the Economy* (Paris: OECD).

—— (1987a) *The Future of Migration* (Paris: OECD).

—— (1987b) *Financing and Delivering Health Care* (Paris: OECD).

—— (1988a) *Ageing Populations: The Social Policy Implications* (Paris: OECD).

—— (1988b) *Reforming Public Pensions* (Paris: OECD).

—— (1990) *Health Care Systems in Transition* (Paris: OECD).

Ogden, P. E. (1989) International migration in the nineteenth and twentieth centuries. In P. E. Ogden and P. E. White (eds) *Migrants in France: Population Mobility in the Later Nineteenth and Twentieth Centuries* (London: Unwin Hyman), pp. 34–59.

Ogden, P. E. and Huss, M.-M. (1982) Demography and pronatalism in France in the nineteenth and twentieth centuries. *Journal of Historical Geography* 8 (3), pp. 283–98.

Ogden, P. E. and White, P. E. (eds) (1989) *Migrants in France: Population Mobility in the Later Nineteenth and Twentieth Centuries* (London: Unwin Hyman).

Olshansky, S. J. (1988) On forecasting mortality. *Milbank Memorial Fund Quarterly* 66, pp. 482–530.

Oppong, C. (1988) The effects of women's position on fertility, family organization and the labour market: some crisis issues. In IUSSP conference on *Women's Position and Demographic Change in the Course of Development* (Oslo: IUSSP), pp. 43–66.

Pennec, S. (1989) La politique familiale en Angleterre-Galles depuis 1945. *Population* 44, pp. 417–28.
Penninx, R. (1986) International migration in Western Europe since 1973: developments, mechanisms and controls. *International Migration Review* 20, pp. 951–72.
Penninx, R. and Muus, P. (1989) No limits for migration after 1992? The lessons of the past and a reconnaissance of the future. *International Migration* 27, pp. 373–88.
Phizaclea, A. (ed.) (1983) *One-Way Ticket: Migration and Female Labour* (London: Routledge).
Pinder, D. (ed.) (1990) *Western Europe: Challenge and Change* (London: Belhaven Press).
Potts, L. (1990) *The World Labour Market. A History of Migration* (London: Zed Books).
Poulain, M. (1990) Une méthodologie pour faciliter la cartographie des niveaux de mortalité en l'absence de donnés sur les décès par âge. *Espace, Populations, Sociétés* 8 (3), pp. 387–91.
Preston, S. H., Keyfitz, N. and Schoen, R. (1972) *Causes of Death: Life Tables for National Populations* (New York: Seminar Press).
Prioux, F. (1989) Fécondité et dimension des familles en Europe occidentale. *Espace, Populations, Sociétés* 7 (2), pp. 161–76.
Pyle, J. (1986) Export-led development and the underemployment of women: the impact of discriminatiory policy in the Republic of Ireland. Paper for ASIDA (New York).
Rallu, J.-L. (1983) Permanence des disparités régionales de la fécondité en Italie. *Population* 38, pp. 29–61.
Rallu, J.-L. and Blum, A. (eds) (1991) *European Population 1. Country Analysis* (London: John Libbey).
Rees, P. H. and Ram, S. (1987) Projections of the residential distribution of an ethnic group: Indians in Bradford. *Environment and Planning A* 19, pp. 1323–58.
Rees, P. H., Stillwell, J. and Convey, A. (1992) Intra-community migration and its impact on the demographic structure at the regional level. Working Paper 92/1, School of Geography, University of Leeds.
Reinhard, M., Armengaud, A. and Dupâquier, J. (1968) *Histoire Générale de la Population Mondiale* (Paris: Montchrestien).
Revue Française des Affaires Sociales (1990) *Politiques Sociales en Europe: Quelles Convergences?* 3 (July–September), pp. 3–180.
Richmond, A. H. (1987) Caribbean immigrants in Britain and Canada: socio-demographic aspects. *Revue Européenne des Migrations Internationales* 3, pp. 129–50.
Rivière, D. (1987) Migration d'entreprises et migration de main-d'oevre en Italie. *Annales de Géographie.*
Rix, S. E. and Fisher, P. (1982) *Retirement Age Policy: An International Perspective* (New York: Pergamon).
Robinson, V. and Hale, S. (1990) Un apprentissage difficile: le programme gouvernemental pour l'installation des réfugiés vietnamiens au Royaume-Uni. *Espace, Populations, Sociétés* 8 (2), pp. 207–20.
Rogers, A. (1989) The elderly mobility transition: growth, concentration

and tempo. *Research on Aging* 11 (1), pp. 3–32.

Rogers, A. and Willekens, F. (eds) (1986) *Migration and Settlement: A Comparative Study* (Dordrecht: Reidel).

Roussel, L. (1987) Deux décennies de mutations démographiques. *Population* 42, pp. 429–48.

Roussel, L. and Festy, P. (1979) *Recent Trends in Attitudes and Behaviour Affecting the Family in Council of Europe Member States* (Strasbourg: Council of Europe).

Ruile, A. (1983) L'intégration des étrangers dans les grandes villes allemandes: bases théoriques, observations empiriques et évaluation d'une situation. *Espace, Populations, Sociétés* 1 (2), pp. 89–102.

Ryder, N. B. (1980) Components of temporal variations in American fertility. In R. W. Hiorns (ed.) *Demographic Patterns in Developed Societies* (London: Taylor and Francis), pp. 11–54.

Salt, J. (1985) The geography of unemployment in the United Kingdom in the 1980s. *Espace, Populations, Sociétés* 3 (2), pp. 349–56.

—— (1988) Highly-skilled international migrants, careers and international labour markets. *Geoforum* 19, pp. 387–99.

—— (1992) Migration processes among the highly skilled in Europe. *International Migration Review* 26, pp. 484–505.

Salt, J. and Clout, H. D. (eds) (1976) *Migration in Post-War Europe: Geographical Essays* (Oxford: Oxford University Press).

Salvini, S. (1984) Fecondità e partecipazione della donna alla vita lavorativa. *Rapporti Monografici* 2, pp. 119–50.

de Sandre, P. (1976) Examen critique des politiques démographiques en Europe, Conseil de l'Europe, Sem./P.S.(76)7F. *Séminaire du Conseil de l'Europe sur l'Incidence d'une Population Stationnaire Décroissante en Europe*, 6–10 September 1976 (Strasbourg: Council of Europe).

Sardon, J. P. (1986) Evolution de la nuptialité et de la divortalité en Europe depuis la fin des années 1960. *Population* 41, pp. 463–82.

Sauvy, A. (1973) Population growth and policy in France. *International Journal of Health Services* 3 (4), pp. 863–7.

Schaffer, H. G. (1981) *Women in the Two Germanies. A Comparative Study of a Socialist and a Non-Socialist Society* (New York: Pergamon Press).

Schmid, J. (1984) *Le Contexte des Tendances Récentes de la Fécondité dans les États Membres du Conseil de l'Europe*, Demographic Studies 15 (Strasbourg: Council of Europe).

Schnapper, D. J. (1976) Tradition culturelle et appartenance sociale: émigrés italiens et migrants français dans la région parisienne. *Revue Française de Sociologie* 17, pp. 485–98.

Schneider, W. H. (1990) *Quality and Quantity. The Quest for Biological Regeneration in Twentieth-Century France* (Cambridge: Cambridge University Press).

Schofield, R. S. and Reher, D. (1991) The decline of mortality in Europe. In R. S. Schofield, D. Reher and A. Bideau (eds) *The Decline of Mortality in Europe* (Oxford: Clarendon Press), pp. 1–17.

Schofield, R. S., Reher, D. and Bideau, A. (eds) (1991) *The Decline of Mortality in Europe* (Oxford: Clarendon Press).

Schoorl, J. J. (1990) Fertility adaptation of Turkish and Moroccan women in the Netherlands, *International Migration* 28, pp. 243–66.

Schwarz, K. (1986) One-parent families in the Federal Republic of Germany. In F. Deven and R. L. Clinquet (eds) *One-Parent Families in Europe: Trends, Experiences, Implications* (The Hague/Brussels: NIDI/CBGS), pp. 141–54.

—— (1988) Household trends in Europe after World War II. In N. Keilman, A. Kuijsten and A. Vossen (eds) *Modelling Household Formation and Dissolution* (Oxford: Clarendon Press), pp. 67–83.

—— (1989) Les effets démographiques de la politiques familiale en RFA et dans ses Länder depuis la Seconde Guerre Mondiale. *Population* 44, pp. 395–415.

Sels, C. and Dumon, W. (1990) Family and population policy in Belgium. In W. Dumon (ed.) *Family Policy in EEC Countries* (Luxembourg: Office for Official Publications of the European Community), pp. 15–32.

Shaw, C. (1988) Latest estimates of ethnic minority populations. *Population Trends* 51, pp. 5–8.

Silver, M. (1965) Births, marriages and business cycles. *Journal of Political Economy* 74, pp. 237–55.

Simons, J. (1986) Culture, economy and reproduction in contemporary Europe. In D. Coleman and R. S. Schofield (eds) *The State of Population Theory* (Oxford: Basil Blackwell), pp. 256–78.

Soren, J. (1990) Danish family policy and the absence of a population policy. In W. Dumon (ed.) *Family Policy in EEC Countries* (Luxembourg: Office for Official Publications of the European Community), pp. 33–57.

Speigner, W. (1990) Models of demographic policy: the GDR case. *Conferenza Popolazione, Società e Politiche Demografiche per l'Europa*, Torino, 4–6 April 1990 (Turin: Fondazione Giovanni Agnelli and the European Foundation for Population Studies).

Sporton, D. (1990) *The Differential Fertility of Immigrants within the Ile-de-France Region*. Unpublished Ph.D. thesis, University of Sheffield.

—— (1991) The differential fertility of immigrants within the Paris region, France. *Kieler Geographische Schriften* 78, pp. 187–202.

Sprink, J. and Hellmann, W. (1989) Finanzielle Belastung oder ökonomisches Potential? Regionale unterschiedliche Konsequenzen der Aussiedlerzustroms. *Informationen zur Raumentwicklung* 5, pp. 323–9.

Stillwell, J., Rees, P. H. and Boden, P. (eds) (1992) *Population Redistribution in Britain. Volume Two of Migration Patterns and Processes* (London: Belhaven Press).

Straubhaar, T. (1988) International labour migration within a Common Market: some aspects of EC experience. *Journal of Common Market Studies* 27, pp. 44–62.

Symeonidou, H. (1990) Family policy in Greece. In W. Dumon (ed.) *Family Policy in EEC Countries* (Luxembourg: Office for Official Publications of the European Community), pp. 125–70.

Tabah, L. (1991) *World Demographic Trends and Consequences for Europe*, Population Studies 20 (Strasbourg: Council of Europe).
—— (n.d.) *L'Intégration des Politiques Démographiques et des Politiques Socio-économiques dans les Pays de l'Europe de l'Est* (Paris).
Tas, R. F. J. (1987) La population originaire du Surinam et des Antilles Néerlandaises aux Pays-Bas. *Revue Européenne des Migrations Internationales* 3, pp. 69–90.
Teitelbaum, M. S. and Winter, J. (1985) *The Fear of Population Decline* (London: Academic Press).
Thane, P. M. (1990) The debate on the declining birth-rate in Britain: the menace of an ageing population, 1920s–1950. *Continuity and Change* 5, pp. 203–30.
Thumerelle, P.-J. (ed.) (1990–91) *Les Inégalites Géographiques de la Mortalité, Espace, Populations, Sociétés* 8 (3), 9 (1), special issues.
Tinacci Mossello, M. (1986) La mobilità territoriale della popolazione in Italia. *Revista Geografica Italiana.*
Tribalat, M. (1987) Evolution de la natalité et de la fécondité des femmes étrangères in RFA. *Population* 42, pp. 370–8.
Uner, S. (1991) The changing structure of the European labour force. In *Proceedings of the Seminar on Present Demographic Trends and Lifestyles in Europe, 18–20 September 1990* (Strasbourg: Council of Europe), pp. 179–208.
United Nations (1973) Families and households. In *Determinants and Consequences of Population Trends* (New York: United Nations), pp. 336–97.
—— (1979) *Manual IX: The Methodology of Measuring the Impact of Family Planning Programs on Fertility*. Population Studies 61 (New York: United Nations).
—— (1985a) *Studies to Enhance the Evaluation of Family Planning Programs*. Population Studies 87 (New York: United Nations).
—— (1985b) *The World Aging Situation: Strategies and Policies*. Department of International Economic Affairs (New York: United Nations).
—— (1989) *Prospects of World Urbanization, 1988*. Population Studies 112 (New York: United Nations).
Valkonen, T. (1987) Social inequality in the face of death. *European Population Conference, Jyväskylä, Finland, 1987* (Liège: IUSSP), pp. 201–61.
Vallin, J. (1984) Politiques de santé et mortalité dans les pays industrialisés. *Espace, Populations, Société* 2 (3), pp. 13–31.
van Amersfoort, H. (1982) *Immigration and the Formation of Minority Groups: The Dutch Experience, 1945–1975* (Cambridge: Cambridge University Press).
—— (1987) Résidence et groupes ethniques dans les villes néderlandaises: classe, race ou culture? *Revue Européenne des Migrations Internationales* 3, pp. 91–116.
van Amersfoort, H. and Surie, B. (1987) Reluctant hosts: immigration into Dutch society, 1970–1985. *Ethnic and Racial Studies* 10, pp. 169–85.

van Amersfoort, J. M. M. and Cortie, C. (1973) Het patroon van de Surinaamse vestiging in Amsterdam in de periode 1968 t/m 1970. *Tijdschrift voor Economische en Sociale Geografie* 64, pp. 283–94.

van de Kaa, D. J. (1987) Europe's second demographic transition. *Population Bulletin* 42 (1), pp. 3–57.

—— (1991) European migration at the end of history. Paper presented at the European Population Conference, Paris, 21–5 October 1991.

van de Walle, E. (1974) *The Female Population of France in the Nineteenth Century: A Reconstruction of Eighty-Two Départements* (Princeton: Princeton University Press).

van den Akker, P. A. M. and van der Avort, A. J. P. M. (1986) Children after parental divorce: short-term and long-term consequences and conditions for adjustment. In F. Deven and R. L. Cliquet (eds) *One-Parent Families in Europe: Trends, Experiences, Implications* (The Hague/Brussels: NIDI/CBGS), pp. 83–110.

van Hoorn, F. J. J. H. (1987) *Onder Anderen. Effekten van de Vestiging van Mediterranen in Naoorlogse Woonwijken.* Nederlandse Geographische Studies 50.

van Poppel, F. (1979) Regional differences in mortality in western and northern Europe: a review of the situation in the seventies. In *Proceedings of the Meeting on Socioeconomic Determinants and Consequences of Mortality, Mexico, 1979* (New York and Geneva: United Nations and World Health Organisation), pp. 251–372.

Vandermotten, C. and Grimmeau, J. (1983) Réflexions épistémologiques pour une géographie de l'emploi/non-emploi. *Espace, Populations, Sociétés* 1 (1), pp. 19–30.

—— and —— (1985) Principales caractéristiques régionales des marchés du travail dans la Communauté Européenne. *Espace, Populations, Sociétés* 3 (2), pp. 441–60.

Vichnevsky, A., Oussova, I. and Vichnevskaia, T. (1991) Les conséquences des changements intervenus à l'est sur les comportements démographiques. Paper presented to the European Population Conference, Paris, 21–5 October 1991.

Vichnevsky, A. and Zayontchkovskaia, J. (1991) L'émigration de l'ex-Union soviétique: prémices et inconnues. *Revue Europénne des Migrations Internatioales* 7, pp. 5–29.

Villeneuve-Gokalp, C. (1990) Du mariage aux unions sans papiers: histoire récente des transformations conjugales. *Population* 45, pp. 265–98.

Vining, D. R. and Kontuly, T. (1978) Population dispersal and major metropolitan regions: an international comparison. *International Regional Science Review* 3, pp. 49–73.

de Vries, J. (1984) *European Urbanization, 1500–1800* (London: Methuen).

Vuddamalay, V. (1990) L'insertion socio-professionnelle chez les nouveaux immigrés. Le cas des Mauriciens en France. *Espace, Populations, Sociétés* 8 (2), pp. 231–9.

Vukovich, G. (1991) *Population Aging in Hungary: Selected Aspects* (Malta: International Institute on Aging).

Wabe, J. S. (1986) The regional impact of deindustrialisation in the European Community. *Regional Studies* 20, pp. 250–75.

Walker, A., Guillemard, A. M. and Alber, J. (eds) (1991) *Social and Economic Policies and Older People* (Brussels: CEC).

Warnes, A. M. (1989a) The ageing of populations. In A. M. Warnes (ed.) *Human Ageing and Later Life* (London: Edward Arnold), pp. 47–66.

—— (1989b) Elderly people in Great Britain: variable projections and characteristics. *Journal of Care of the Elderly* 1, pp. 7–10.

—— (1992a) Demographic processes and health forecasts. In R. J. Newman (ed.) *Orthogeriatrics: Comprehensive Orthopaedic Care for the Elderly Patient* (Oxford: Butterworth-Heinemann), pp. 1–12.

—— (1992b) Population characteristics and trends. In A. M. Warnes (ed.) *Homes and Travel: The Domestic and Local Setting of Third Age Lives* (London: Carnegie Foundation).

Watkins, S. C. (1990) From local to national communities: the transformation of demographic regimes in Western Europe, 1870–1960. *Population and Development Review* 16, pp. 241–72.

—— (1991) *From Provinces into Nations: Demographic Integration in Western Europe, 1870–1960* (Princeton: Princeton University Press).

Wattelar, C. (1982) *Examen Général de la Situation et des Tendances Démographiques en Europe, Département de Démographie, Université Catholique de Louvain, Document de Recherche*, 61.

Weber, E. (1991) Fertility differences in the European socialist countries. *Kieler Geographische Schriften* 78, pp. 229–45.

Werheke, D. (ed.) (1983) *Microelectronics and Women's Jobs* (Geneva: International Labour Office).

Werner, B. (1988) Fertility trends in the UK and in thirteen other developed countries, 1966–86. *Population Trends* 51, pp. 18–24.

Werner, H. (1986) Post-war labour migration in Western Europe: an overview. *International Migration* 24, pp. 543–57.

White, P. E. (1986) International migration in the 1970s: revolution or evolution?. In A. M. Findlay and P. E. White (eds) *West European Population Change* (London: Croom Helm), pp. 50–80.

—— (1987) The migrant experience in Paris. In G. Glebe and J. O'Loughlin (eds) *Foreign Minorities in Continental European Cities* (Wiesbaden: Steiner), pp. 184–98.

—— (1988) Skilled international migrants and urban structure in Western Europe. *Geoforum* 19, pp. 411–22.

White, P. E. and Kesteloot, C. (1990) Les migrations internationales en Europe Occidentale durant les années quatre-vingts. *Espace, Populations, Sociétés* 8 (2), pp. 316–23.

White, P. E. and Woods, R. I. (1983) Migration and the formation of ethnic minorities. *Journal of Biosocial Science*, supplement 8, pp. 7–24.

Wicks, M. and Chester, R. (1990) Family policy and population policy – the UK situation. In W. Dumon (ed.) *Family Policy in EEC Countries* (Luxembourg: Office for Official Publications of the European Community), pp. 103–24.

Wihtol de Wenden, C. (1985) L'immigration italienne en France. I. La

formation et la mobilité. *Studi Emigrazione* 78, pp. 213–25.
Wilkinson, R. G. (ed.) (1986) *Class and Health. Research and Longitudinal Data* (London: Tavistock Press).
Willekens, F. (1988) A life-course perspective on household dynamics. In N. Keilman, A. Kuijsten, and A. Vossen (eds) *Modelling Household Formation and Dissolution* (Oxford: Clarendon Press), pp. 87–107.
Wilson, C. and Woods, R. I. (1991) Fertility in England: a long-term perspective. *Population Studies* 46, pp. 399–415.
Wilson, T. and Wilson, D. (1991) *The State and Social Welfare: The Objectives of Policy* (London: Longman).
Winter, J. M. (1985) *The Great War and the British People* (London: Macmillan).
Woods, R. I. (1987) Approaches to the fertility transition in Victorian England. *Population Studies* 41, pp. 283–311.
—— (1992) *The Population of Britain in the Nineteenth Century* (London: Macmillan).
World Bank (1990) *World Development Report, 1990: Poverty* (Oxford: Oxford University Press).
Zimmerman, J. (ed.) (1983) *The Technological Women* (New York: Praeger).

# Index